How to Get Started With Modems

by Jim Kimble

Editing by Tina Berke and Gretchen Lingham
Cover illustration by Randy Verougstraete
Book design by Collette Murphy
Art direction by Kay Thorogood

Computer Publishing Enterprises
P.O. Box 23478
San Diego, CA 92123

0-945776-05-5

CONTENTS

FOREWORD

More and more people are enjoying the fun and convenience of home computers, yet only a small percentage of computer hobbyists ever discover the power and utility of a modem. This sad situation is probably due in part to the frightening media hype surrounding modems, modeming and modemers; in part to the widely shared perception that modems and communications software are hard to connect, figure out and use.

This book sets out to clear the fog surrounding telecommunications. Time and time again people have asked for help with setting up their modems, communications software or BBSs. Yet compared with the early days of computing, today's modems and communications software are well-designed and surprisingly easy to use.

Jim has tried to make it all even more accessible by explaining every detail of the telecommunications process: from the phone lines to the host computers, from the software to commercial BBSs. With the clear, understandable descriptions of what's going on every step of the way, you'll be a telecommunications expert in no time. The glossary and index in the back will make you want to keep this book handy on a shelf near your computer. And the appendices will detail sources for communications software, installing a modem, the Hayes command set and more.

So sit down in front of your computer, turn the page, and get started with modems today!

Tina Berke
July 1989

INTRODUCTION

Are You Ready for a Modem?

When you reach the point in your life when you'd rather spend time with the family than sit down to an 18-hour session of intense computer use, it's time for an addition to your collection of high-tech hardware. It's time for an addition that will not only extend your computer's reach beyond the confines of your desktop, but zap many, many hours out of your already busy schedule. It's time for a modem.

ABOUT THIS BOOK

The communications side of computing is by far the most complicated. There are a lot of buzz words involved in buying a new computer, but there are nearly twice as many involved in getting started with modems and telecommunications.

In reading this book, you have an advantage many others do not. You have the opportunity to draw on the knowledge and experiences (both good and bad) of someone who has already taken the proverbial "plunge."

This book was written in a style which assumes you'll read it like a novel—front to back. Each chapter will build upon the previous one. As you follow along, you'll learn how to purchase your first modem, install it on your own, and use it like a pro. There is a glossary in the back in case you need a more detailed explanation of a particular word or term. However, you'll probably never have to use it.

Each of the confusing, technical aspects of telecommunications are dissected here and explained in plain and simple English. When you reach this book's conclusion, you will be armed with enough information to perform verbal battle with even the *fiercest* of modem salespeople. You will be a Modem Guru in Training.

PART ONE

CHAPTER 1

What Is Data Communications?

The phrase "data communications" refers to two computers electronically "talking" to one another. For personal computers, this electronic conversation is made possible through the use of modems. Modems are nifty little devices that allow your computer to communicate with other computers over ordinary phone lines. Your particular brand of computer makes no difference; you can be the proud owner of a shy, little Apple II+ or a hulking Cray Y/MP supercomputer and still be able to slap a modem on a spare serial port and join the exciting world of telecommunications.

The word modem is a contraction of MOdulator-DEModulator. A modem basically works by taking electrical signals from your computer, translating them into audible tones, and broadcasting them along ordinary phone lines to another computer/modem setup. That's the modulator part. The demodulator side of a modem receives tones from another computer/modem setup, translates them into electrical signals, and gives them to your computer.

Even though the computers are conceptually talking to each other, we humans are concerned only with the content of their conversation. How the two machines conduct their dialogue is entirely their own business. Because we leave our modems alone, they return the favor by paying no attention to what's being sent through them—you can transfer text documents, software programs, and even your favorite game's high-score files without a hitch.

HOW IT WORKS

When talking to the computer on the other end of the phone line, you're doing more than just chit-chatting. You're actually in command of the other system. When you type the letter "G" on your keyboard, your computer does nothing more than send that character to the other machine (via the modem). When the other machine receives it, it uses the "G" exactly as if you had typed it from *that* computer's keyboard; the receiving system ignores the fact that the "G" was sent from another machine.

If this sounds really confusing, that's because it is. It's hard enough figuring out how to use your own computer, much less someone else's. But luckily the modems and computers take care of all the dirty work. Once you set everything up, all you'll have to do is tap out a few keystrokes to start things rolling. Very rarely does anyone ever concern themselves with the low-level, morbid details of data communications ... unless they're really weird.

MOODY CONVERSATIONALISTS

Modems are very moody beasts. They can't just hold a normal conversation with another of their species; instead, they have to use all kinds of technical-sounding words to describe the way they're chatting. Let's look at some of these technical terms.

Duplex

Picture two kids sitting in separate treehouses; we'll call one of them Jim and the other Dan. Jim and Dan need to quietly pass messages between one another, but have found climbing up and down a tree to pass a verbal or written message a little impractical.

***Figure 1:** Here are Jim and Dan illustrating the process of characters being sent one at a time in one direction. This is the duplex method of data communications.*

But, being only kids, they can't afford keen electronic toys such as walkie-talkies. What can they do?

Being the energetic youths they are, they come up with the idea of stringing a single wire between the two treehouses, and sliding beads to one another. Each of the beads has a single letter painted on it. For example, Dan slides a bead with the letter "H" to Jim. Then he slides an "E," "L," "L" and "O" down the wire—until finally he forms the word "HELLO." Because Dan has all the beads, Jim can only receive messages, and Dan can only send them.

To make for two-way communication, Jim and Dan whip up some more beads so Jim can respond to Dan's messages. Now they can send beads down the wire in either direction, but what happens if Jim and Dan both were to send a bead at the same time? The beads would collide and get stuck in the middle of the wire.

The solution is to string up a second wire: one wire to send beads and one wire to receive them. The wires are reversed on each end; Jim's send wire would be Dan's receive wire and vice versa. Now they can send messages back and forth all day long.

In data communications, these three scenarios represent "simplex," "half duplex" and "full duplex," respectively. Simplex is where information can be sent in one direction only, as in beads down a single wire. Half duplex is where information can be sent in both directions, but only one way at a time. Full duplex is where there are two wires, one for sending and one for receiving, both at the same time.

The wires being used in our example are the equivalent of a "carrier signal" sent by a modem over a phone line. The carrier signal is a tone transmitted by a modem that tells the other modem, "Hey! I'm still listening!" When a modem "drops carrier," it means the other computer has stopped emitting the carrier signal and is no longer listening.

The beads are the equivalent of data bits being transmitted between the two computers. The bits of data "ride" on the carrier signal, just like beads on a wire. The two carrier signals are distinguished by two different tone pitches. The pitches are referred to as "originate" and "answer." The modem can determine who sent what by examining the carrier pitch. If your modem was set to the originate pitch, any characters found on the answer pitch would be coming from the other computer.

Generally speaking (dangerous when it comes to computers), almost everybody uses full duplex for data communications. And generally, you should set your modem to originate mode when calling another computer. When a different computer initiates the connection, you would normally set your modem to answer mode. In most cases, this minor (although important) detail is handled by your communications software, as we'll see later.

HERE, HAVE A BYTE

Modems can't transmit complete characters all at once, as our "beads on a wire" analogy suggests. Instead, they have to break each individual character, called a byte, into a number of "bits." The term bits is short for "binary digit," and *ten* bits make up a byte. (Yes, ten! Your computer uses eight bits to make a byte, but your modem tacks two extra bits on each byte for its own use.)

To imagine how this works, think of each byte as a stack of eight beads. For Jim to send these eight beads to Dan, he'd have to un-stack them, slide them along a wire (from our earlier example), and have Dan restack them on the other end.

In computer communications, these beads are bits. There are two kinds of bits: on-bits and off-bits. An on-bit is a one, and an off-bit is a zero (nothing). These bits are arranged in sets of eight to represent characters inside the computer.

For example, the eight bits "01000001" represent the letter "A." The eight bits "01000010" just happen to represent the letter "B." There are 256 combinations of eight bits in a byte, and each one of these combinations is associated with a letter of the alphabet, a number, a punctuation symbol or a variety of symbols having internal meaning only to the computer.

Luckily, the first half of these bit combinations, numbered zero through 127, is standardized on most computers by the American Standard Code for Information Interchange. These ASCII characters form a common language by which text can be transmitted between computers.

You can send the letter "A" (01000001 in bit-talk) from your IBM to an Apple or Commodore computer because all of these systems recognize the byte "01000001" as the letter "A." All three of these computers use the ASCII scheme to represent characters.

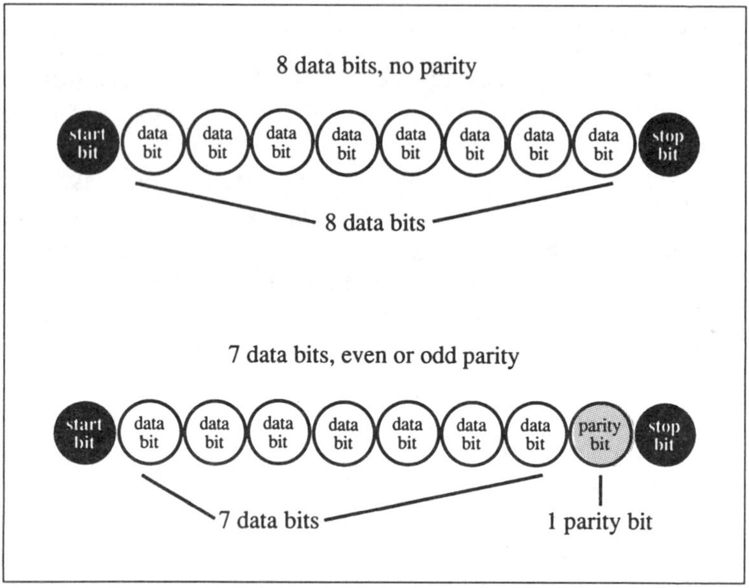

***Figure 2:** Start and stop bits are always "1." The value of the parity bit depends on the byte transmitted and whether you want even or odd parity.*

TRANSMISSION FORMAT

As you can imagine, it would be hard to keep track of all those beads being slid down our imaginary wire. To help maintain some kind of order, in addition to data bits, the modem uses three extra bits: a stop bit, a start bit and an optional parity bit.

To illustrate how each of these work, let's say we're going to send the letter "A" down the wire. To help Dan at the other end, we're going to send a special bead that indicates we're going to start sending a new character. This bead is called a start bead. We *always* send a start bead before transmitting a new character.

When Dan sees the start bead, he knows he's going to begin a new stack of on- and off-beads. After receiving a combination of eight on-beads and off-beads and stacking them in order, Dan is sent a stop bead that indicates we're done with the stack. The modem performs these exact steps—with bits instead of beads.

ASCII CHART											
DEC	HEX	CHAR	DEC	HEX	CHAR	DEC	HEX	CHAR	DEC	HEX	CHAR
00	00	NUL	32	20	SP	64	40	@	96	60	`
01	01	SOH	33	21	!	65	41	A	97	61	a
02	02	STX	34	22	"	66	42	B	98	62	b
03	03	ETX	35	23	#	67	43	C	99	63	c
04	04	EOT	36	24	$	68	44	D	100	64	d
05	05	ENQ	37	25	%	69	45	E	101	65	e
06	06	ACK	38	26	&	70	46	F	102	66	f
07	07	BEL	39	27	'	71	47	G	103	67	g
08	08	BS	40	28	(	72	48	H	104	68	h
09	09	HT	41	29	)	73	49	I	105	69	i
10	0A	LF	42	2A	*	74	4A	J	106	6A	j
11	0B	VT	43	2B	+	75	4B	K	107	6B	k
12	0C	FF	44	2C	,	76	4C	L	108	6C	l
13	0D	CR	45	2D	–	77	4D	M	109	6D	m
14	0E	SO	46	2E	.	78	4E	N	110	6E	n
15	0F	SI	47	2F	/	79	4F	O	111	6F	o
16	10	DLE	48	30	0	80	50	P	112	70	p
17	11	XON	49	31	1	81	51	Q	113	71	q
18	12	DC2	50	32	2	82	52	R	114	72	r
19	13	XOFF	51	33	3	83	53	S	115	73	s
20	14	DC4	52	34	4	84	54	T	116	74	t
21	15	NAK	53	35	5	85	55	U	117	75	u
22	16	SYN	54	36	6	86	56	V	118	76	v
23	17	ETB	55	37	7	87	57	W	119	77	w
24	18	CAN	56	38	8	88	58	X	120	78	x
25	19	EM	57	39	9	89	59	Y	121	79	y
26	1A	SUB	58	3A	:	90	5A	Z	122	7A	z
27	1B	ESC	59	3B	;	91	5B	[	123	7B	{
28	1C	FS	60	3C	<	92	5C	\	124	7C	|
29	1D	GS	61	3D	=	93	5D	]	125	7D	}
30	1E	RS	62	3E	>	94	5E	^	126	7E	~
31	1F	US	63	3F	?	95	5F	_	127	7F	DEL

***Figure 3:** The ASCII chart above shows the common 128-character set used in telecommunications. This chart isn't something that most people memorize, but it is a handy reference for some of the codes used in applications and data files.*

THE PARITY BIT

The parity bead—er—bit ranks high on the list of Least Understood Aspects of Telecommunications. Simplified, the parity bit is just a crude form of error checking.

In our earlier example of start bits and stop bits, we used eight data bits (those ones and zeros) for each character. Because each telecommunications character can consist of only ten bits, you have to sacrifice a data bit when an optional parity bit is added to the picture. That means our new *word format* consists of one stop bit, *seven* data bits, a parity bit and a stop bit. Luckily, the entire ASCII table, consisting of 127 characters, appears as combinations of seven ones and zeros.

How Does it Work?

The parity bit, like a data bit, is either a one or a zero; it's used for very simple error checking. There are two basic types of parity: even and odd. When communicating at *even* parity, your modem evaluates the data bits and counts the number of ones making up a single character. If the number of ones is odd, the parity bit is a "1," to make the total number of ones even. If the number of ones is an even number, the parity bit is a "0" to keep the total number of ones even.

The receiving modem dissects each incoming character and checks to make sure that there is an even number of ones making up each byte. If you're using even parity and the number of ones making up a single byte is not an even number, the receiving computer will indicate that it has received a parity error. This error condition makes the user aware of a problem with the communications link.

Odd parity is exactly like even parity, except the modem is working to make the number of ones in each byte an odd number. Typically, very few computers utilize the parity bit—99 percent of all bulletin boards in the free world use eight data bits and no parity. If you do happen across a really bizarre computer system which requires a parity bit, it will more than likely be operating at seven data bits and *even* parity.

CHAPTER SUMMARY

This wraps up the basics of telecommunications. If you've completely understood everything in Chapter One, you now have a technical background in data communications probably rivalling that of most other modem users. None of the remaining discussions and examples in this book will get any more technical than the chapter you've just read.

CHAPTER 2

Modems

Buying a new modem is a lot like buying a new car (except it's easier to finance a new modem). Cars come with lots of neato features and luxuries to make your life easier. So do modems. As with air conditioning and rear-window defrosters on cars, extra features on modems will cost extra money. In this chapter I'll walk you through all the things your modem *should* do, and all the things your modem *could* do. When you go out pricing your new investment, you can decide if the extra goodies are worth the extra money.

INDUSTRY RANKINGS

Modems are rated in a number of categories, the main one being speed. A modem's speed is its "baud rate" (pronounced bod rate). The common speeds are 300 (extremely slow and almost obsolete), 1200 (acceptable), 2400 (fast) and 9600 (very fast and very expensive).

People will often brag about their computers by saying something foolish like "I have a 2400 baud modem." In reality, the person doesn't have a 2400 baud modem. What he or she really has is a 2400 bits per second (bps) modem. Most people mistakenly consider baud and bps to be the same thing; they're not. However, because the terms are so often and so blatantly confused, you might as well join the ranks and simply consider bps and baud to mean the same thing.

The confusion all began because users failed to realize that baud describes one aspect of the modem, and bps describes another. A complete description of baud rates would be long, boring, and well beyond the scope of this book, so I'll skip it. On the other hand, the concept of bps is important, and a heck of a lot easier to swallow.

Bits per Second

At 300 bps, your modem transmits 300 bits every second. Because there are ten bits in a telecommunications byte (as discussed in Chapter One), 300 bps translates to approximately 30 bytes a second. Here's why: 300 bps divided by ten bits in a byte equals 30 bytes per second.

Along the same lines, at 1200 bps your modem is able to transmit up to 120 bytes every second. The same would follow for 2400 bps and 9600 bps.

The faster modems are nice because they can transmit more data in less time, but they are also more expensive. To justify spending the extra money, ask yourself what you intend to do with your modem. Are you going to be transferring a lot of big files? If so, a 9600 bps modem might be something you'll want to look into. The faster modems will transmit data faster, so you'll be spending less time on the phone; that could translate into money saved if you're calling computer systems outside your local toll-free area.

However, if you only plan on tinkering around with the local electronic bulletin board systems by sending and receiving relatively short messages, you might think about saving a few bucks by sticking with a cheaper modem. For most humans, 1200 bps is a good reading speed; anything faster would require an Evelyn Woods™ speed-reading course. Unfortunately, the first time you see a 2400 bps modem in action, you'll wish you had one. Face it, they're nice. It's a decision you and your pocketbook are going to have to make.

An advantage to purchasing a high-end modem is that it's "downward compatible." This means a 2400 bps modem will talk to not only every other 2400 bps modem in the world, but also every 1200 bps and every 300 bps modem. To carry this even further, a 1200 bps modem will talk to other 1200 bps modems *and* every 300 bps modem. This is accomplished by a ritual called "handshaking." When two modems first connect with one another, they spend a couple milliseconds letting each other know a little bit about themselves. Each of the modems lets the other know the fastest speed at which it's capable of transmitting. The two then agree on the fastest speed they can both use.

For example, pretend your friend with a 1200 bps modem calls your SuperMondo 2400 bps modem. Your modem answers the phone and says, "I can talk at 300 bps, 1200 bps or 2400 bps." Your friend's modem says, "That's great, but I can only talk at 300 bps or 1200 bps." The two then agree that 1200 bps is the fastest they can *both* understand, so the faster of the two (your modem) automatically drops down to 1200 bps and resumes the connection. This is completely automatic to the user; all the dirty work is done by the hardware (the way it should be).

The point you should be grasping here is that a modem can transmit at every speed *up to its rated speed*. A 2400 bps modem can speak at zero to 2400 bps, a 1200 bps modem can speak at zero to 1200 bps, and a 300 bps modem can speak at zero to 300 bps.

The 9600 bps modems are incredibly fast, but they're also incredibly expensive.

INSIDE, OUTSIDE, UPSIDE DOWN

Regardless of the modem's brand, its speed or any other factor, it falls in one of two categories: either *external* or *internal*.

Internal Modems

Internal modems live *inside* your computer. They plug into an open slot just like any other card. Internal modems are usually less expensive because they don't need things like a pretty case or an external power supply (they steal a little bit of juice from your computer, but that's okay, because your computer has plenty to spare). Because they plug directly into your computer, they also don't require any special cables. Most importantly, they don't take up any valuable desk space. All you do is plug in a phone cord that goes from your wall to the modem and you're ready to terrorize computer systems worldwide.

Figure 4: *The funny-looking gadget on the left is an internal modem. As you can see, there is not much to it, but they can accomplish everything an external modem can. On the right the external modem looks a lot more complete, but remember, half the cost goes into the case and the little lights.*

External Modems

External modems, as you have probably guessed, sit outside your computer. These beasts tend to be more expensive than internal modems (anywhere from $20 to $100 more, depending on the brand and the dealer), but their advantages far outweigh the disadvantage of the beefier price tag.

External modems are equipped with a row of diagnostic lights that give the user an idea of just what the modem is doing at any given time. The lights are usually labeled HS (High Speed), AA (Auto Answer), CD (Carrier Detect), OH (Off Hook), RD (Receive Data), SD (Send Data), TR (Terminal Ready), MR (Modem Ready), RS (Request to Send), CS (Clear to Send) and AL (Analog Loopback). You "read" these lights by noting if they're illuminated or not. For example, if the OH lamp is on, you know the modem has the phone off the hook; if the SD lamp is lit, you can tell at a glance that your modem is sending data; if the CD light is on, your modem has detected another modem's carrier signal; and so on.

Because an internal modem is inside your computer's case, you don't have the advantage of these diagnostic lights. By using the lamps, you can tell at a quick glance if any communications problems you may be suffering are the fault of *your* computer or modem, or the fault of the computer/modem setup on the other end of the phone line.

Price Versus Performance

Regardless of how good a deal you may be able to finagle, do not purchase a 300 bps modem. The 300 bps modems were considered a good thing just a few years ago, but as the price of the faster modems has dramatically decreased, the 300 bps standard has quickly fallen by the wayside. In fact, 300 bps modems are on the verge of being classed as obsolete.

The 1200 bps modems are good starters, but even these fast babies are quickly losing their popularity. A relatively well-equipped 1200 bps box can be purchased for less than $150. However, if you can afford it, a 2400 bps modem is a great way to go; they're currently the most popular modems on the market. You'll be able to transfer software programs at exceptional rates of speed, and the text will scroll across your screen faster than you can ever read it.

External modems are also portable. If you had an Apple II+ computer with an internal modem and you decided to upgrade to an IBM clone, your modem wouldn't work with your new investment. The internal Apple modems have to stay inside of Apple machines; IBM internal modems must stay inside of IBM machines. However, if you had an external modem plugged into your Apple II+, you would need nothing more than a new cable to connect the modem to its new boss: the IBM. Unfortunately, this "portability factor" means your modem is going to have to monopolize a serial port on your computer. But by being more portable, it's easier to unload a used external modem versus an internal modem, should you ever be inclined to do so. (Although be warned: Modems don't hold their resale value.)

Hayes Compatibility

No matter what area of the computer industry, compatibility is the buzzword of choice among computer professionals and hobbyists. Modems are no different from the rest of the industry (although many of their users can be called different).

Back in the early days of telecommunications (during the early 1970s), a company called Hayes joined the modem market and produced a new type of modem outfitted with something called the AT (pronounced "Aye - Tee") Command Set. Basically this was a specialized instruction set used to control your modem.

In the AT command set, every command sent to the modem is preceded with the letters "AT," which explains the name. AT means "ATTENTION" to your modem. Through effective use of the AT Command Set, you can tell your modem to do anything from immediately dialing a certain phone number to answering any incoming calls after the 203rd ring. (See Appendix C for more detail.)

Because there wasn't much competition during those early modem days, the Hayes AT Command Set met with extreme popularity. The command set was so popular, in fact, that other companies began to include it in their brands of modems. To this day, the Hayes AT Command Set is a highly sought-after feature within modems; its popularity is so great that most of the communications programs on the market were written with the assumption that the user owned a Hayes compatible modem. After all, what kind of a person would be foolish enough to own a non-Hayes compatible modem?

Owning an "incompatible" modem isn't as bad as having some sort of communicable disease, but it's close. Without Hayes compatibility, you'll still be able to use your modem, but you will almost assuredly lose out on some of the more powerful functions built into your communications software package.

When purchasing a new modem, make sure it's Hayes compatible. And if your hardware dealer tries to tell you Hayes compatibility refers to a modem's ability to talk to another modem, point your finger directly in the guy's face and laugh loudly. You know better than that.

ERROR CORRECTION (MNP)

As discussed in Chapter One, modems translate data into all sorts of peculiar sounds—and they translate all sorts of peculiar sounds coming from another modem into data. When stray clicks, echoes, and sounds leak into the phone line, your modem goes absolutely nuts.

Have you ever made a long distance phone call and happened across a bad connection? Sometimes you can hear your conversation echo in the background, or a loud hissing, or maybe you can faintly hear conversations from other phone lines. Being a human, you can easily disregard this extra noise and simply call Ma Bell obscene names under your breath. Modems, however, can't distinguish these "extra" noises from the actual tones transmitted as data, so the modems makes every attempt to interpret the stray noises into data characters. The result is a screen full of random characters slipped in among the real data.

There is a simple solution to this problem (called line noise). A neat option called MNP (Microcomm Networking Protocol) has been installed in some top-rated modems. MNP works to alleviate line noise by verifying each and every character sent and received by the two respective modems. Here's a simplified explanation of how it works:

The sending modem transmits a character and then sits and waits for the receiving modem to send back the character (as a form of verification). When the receiving modem passes the character back to the sending modem, a positive or negative acknowledgement is transmitted. If the receiving modem received the proper character, a positive acknowledgement is sent and the process starts all over again with the next character. If the receiving modem did *not* receive the proper character, the sending modem will issue a negative acknowledgement along with the original character. The receiving modem discards the character received (because it was obviously in error), and waits for the byte to be retransmitted.

This basic theory is applied on a much higher level to keep the transmission overhead down. MNP is amazingly effective and yet you can't see a noticeable drop in transmission speed while it's active. The disadvantage to the package is price; MNP will add quite a few bucks to the final purchase price of your new modem.

Before you rush out and buy an MNP-equipped modem, be aware that in order for it to work, you must be talking to another modem equally outfitted with MNP. Because they're expensive, MNP-equipped modems tend to be few and far between. When two modems first connect and they begin exchanging salutations and technical information about one another, one of the things they talk about is MNP; if equipped with MNP, the modem that originated the call will query the other modem to see if it's similarly equipped. If so, MNP is used; if not, MNP is turned off and ignored.

DATA COMPRESSION

Another neat feature found in some of the more expensive modems is data compression. Here's how it works:

Two modems equally outfitted with data compression capabilities connect to one another and begin exchanging information. Instead of sending data on a character-by-character basis, the modem saves up a number of characters and then squeezes the characters together. Once the squeezed package of data is ready, the entire thing is transmitted, where it's received and automatically un-squeezed on the other end.

Data compression can increase your throughput by an almost exponential factor. Throughput is the actual number of characters transmitted by your modem every second; while you may have a 2400 bps modem and would expect to see 240 characters sent (or received) every second, line noise and other factors can bring down the transmission rate.

Through the miracle of data compression, some modems are capable of transmitting data over 100 percent faster than their rated speed. For example, one hardware manufacturer sells a 2400 bps modem that transfers the equivalent of 4800 bps when data compression is active!

As with MNP, data compression will mean a slightly higher price tag—and you must be connected to another modem that also has data compression in order for it to work. At the time of this writing, this pretty much means a 9600 bps modem.

PROGRAMMABILITY

One of the most sought-after features in modems is programmability. Modems that boast 100 percent Hayes compatibility should allow you to program what's called an "S-Register."

S-Registers are simply a series of memory locations in your modem that contain various parameters. For example, your modem will answer the phone after the number of rings stored in

S-Register number 1 (noted as S01). All you have to do is tell your modem to store the number of rings (usually between one and 255) you want it to wait before picking up the phone. Other programmable features include the number of milliseconds the modem should wait between the time it answers the phone and the time it responds, and the sequence of characters used to tell the modem to hang up.

If you ever plan on running a BBS using your modem, programmability is something you absolutely *must* have. However, even if you only plan on calling local bulletin boards, be sure your modem has this important feature.

ADDITIONAL FEATURES

Each modem manufacturer tries to cram as many extra features as possible into their box to make it appear better than any of the others. However, in reality, you'll never have the opportunity to exploit many of these features—as your communications software will handle that sort of stuff for you.

For example, some modems offer you the ability to store often-called phone numbers in NVRAM (Non-Volatile RAM, fed by a battery so memory isn't lost when you turn off the power). This may seem like a nice feature, but the task of storing phone numbers is usually performed a little more reliably by your communications software—and your communications software can do more with the numbers stored than your modem (as you'll see in Chapter Three).

Another common feature is a built-in automatic redialer. When activated, the redialer will instruct the modem to continuously dial a phone number until it successfully connects with another modem. This is a handy little feature when you're trying to call a computer bulletin board system that seems constantly busy, but it's something your communications software can fake.

I'm not trying to frighten you away from any of these little features; I just want to make sure you don't believe they're as essential as the manufacturer or the dealer says.

PART TWO

CHAPTER 3

Communications Software

Before you can use your new modem you'll have to find a good comunications software package. These packages range in cost from free (public domain) to $200 or more. Unlike the real world, you *don't* get what you pay for when it comes to communications programs; most of the shareware applications far exceed the expensive commercial programs in quality and features offered.

In the early days of modems and BBSs, when there were very few communications applications available on the market, many of the world's basement programmers wound up writing their own personal communications programs, releasing them as either public domain software or shareware. This is great news for anyone just entering the telecommunications world because this means you can pick up and sample a few communications programs for free. After you've had a chance to play with each of them, you can compile a list of the features you like and begin your quest for the *perfect* communications program.

WHAT DOES THE SOFTWARE DO?

When you connect your computer and modem to one another, the two may seem like a perfect couple—but neither really knows the other exists. This is where communications software steps in.

Communications software acts like a party host—it effectively introduces each one to the other.

While the communications software is running, everything you type into your keyboard (with the exception of a few special function keys) is "trapped" by the communications software and passed *directly* to your modem. Your personal computer doesn't know, and really doesn't care, what you just typed.

The software works in both directions; it also traps anything coming from your modem and displays it directly to your screen. Your personal computer completely ignores all of the incoming characters.

Nearly all the popular communications packages allow you to do neato things other than just directly conversing with your modem. Generally, the more features packed into these programs, the better. Communications programs are simple enough that even novice users can figure them out after reading the documentation, yet the programs are sophisticated enough to bring confusion to anyone brave enough to explore every feature offered.

THE PERFECT COMMUNICATIONS PROGRAM

It's impossible to tell you exactly what communications program is perfect for your needs, but here are a few of the features you'll want to make sure your final selection contains:

Automatic Dialing Directory

Automatic dialing directories store the names, phone numbers, vital statistics and settings for each of your favorite bulletin boards. To instruct the software to dial a bulletin board, all you need do is enter the directory, then enter the number assigned to the BBS in the directory (usually one to 100). The software and speed settings will automatically be set and the number will be dialed.

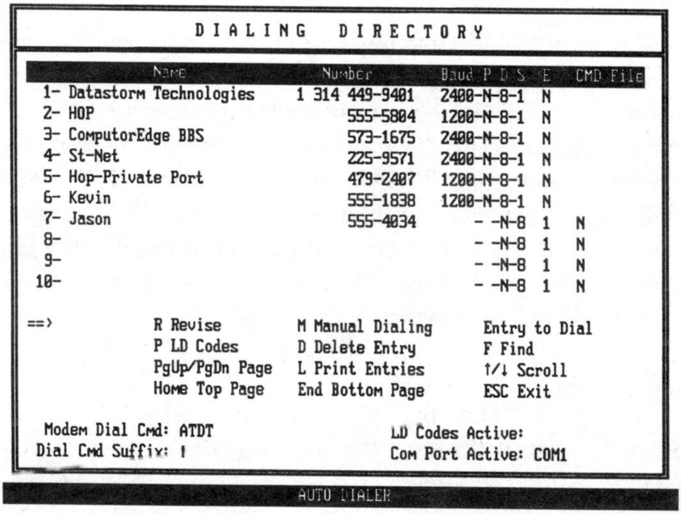

Figure 5: *This is a good example of a dialing directory from* Procomm, *a popular telecommunications program. Notice that all of the settings (baud rate, etc.) can be specifically preset for each listing.*

Automatic Redialer

This little gem usually goes hand-in-hand with an automatic dialing directory (see above). An automatic redialer will continually dial a BBS's phone number until the modem is able to connect. If after dialing a number the modem finds the phone line busy, the modem will hang up and dial the number again. If the line is still busy, the process repeats itself until the BBS eventually answers.

Automatic redialers are a great time saver. As an electronic bulletin board becomes increasingly popular, the system's single phone line is occupied more and more. With an automatic redialer, you can have your computer do all the work involved in calling the BBS while you do more productive things ... such as shooting rubber bands at your cat.

High Speed Compatibility

Some of the older communications programs were written in the days when 1200 bps seemed like an outrageously high speed—2400 bps and 9600 bps modems weren't expected to appear on the consumer market within ten years. Consequently, some of the programs' authors didn't bother to include speeds like 2400 bps or 9600 bps as an option in their software. While it is pretty rare to find a program that doesn't allow the software to address modems at speeds of at least 9600 bps, this is still something to watch out for. If your program boasts of being able to speak to modems at 19,200 bps (sometimes referred to as *19.2KB*), you're doing just fine.

Multiple Protocols

There are probably 50 different communications protocols used for transferring programs (see *Uploads and Downloads*). A good communications program will offer a number of protocols from which to choose when performing transfers. Some of the popular protocols include X-Modem (sometimes called *Christianson Protocol*), Y-Modem, Y-Modem Batch, Z-Modem and Kermit. At the very minimum, make sure your program includes X-Modem as an option. You'll find any additional protocols included with the software useful as you become more of a modeming expert. When it comes to protocols, you'll end up only using one or two of them on a regular basis. However, the more protocols your communications program offers, the more versatile you'll be.

Log Files

A Log File captures every character, incoming and outgoing, from your modem, and copies each of them to a plain ASCII text file. This allows you to save electronic mail messages and other pertinent information on your disk drive for later review. This is especially useful when you call a long-distance bulletin board because you can dump all your awaiting messages to a Log File without spending time reading them on-line. You can then hang up the phone (thus saving in phone bills) and read the messages at your convenience. Once read, you can call the bulletin board again and write your replies.

Printer Logging

When you activate the Printer Log, every piece of text coming in and out of the modem is copied to the printer. This allows you to create a verbatim transcript of your entire on-line session. You can usually toggle the "printer dumping" on and off by pressing just a single function key—so you can capture only the information in which you're really interested. The printer log functions are ordinarily very generic and work with just about every printer around. Some programs even allow you to dump a screen image to the printer. This is like taking a snapshot of the screen and having it developed on the printer, graphics and all.

Login Scripts

Login Scripts allow you to have your computer "login" for you by automatically entering your name and password. You write a miniature program in a super-simple language that is interpreted by the communications program. All your login script basically does is tell the terminal program to wait for a specific string of characters (such as "enter your name:" or "account") and reply with another string (your name or account number). The script "language" is incredibly easy to use and requires no previous programming knowledge whatsoever.

After you become comfortable with some of the full-featured communications programs' script languages, you can have your terminal program not only enter your name and password for you, but also automatically check your mailbox for new messages. Or you could have your login script automatically scan for new public messages, or automatically start downloading a new program, etc. The possibilities are almost endless.

```
%I"First Name?"
%T"Jim!"
%I"Last Name?"
%T"Kimble!"
%I"Is this correct?"
%T"Y!"

%I"Password?"
%T"MYPASSWORD!"
```

Figure 6: *This is a typical example of a login script on a Bulletin Board System. The questions would be a bit more extensive if it were your first time logging on.*

If you happen to call a long-distance bulletin board, you can have your login script automatically dial the phone, enter your name and password, read any new messages, dump them to your printer, and then hang up. Since your communications program can do things much faster than you, using scripts can save you a considerable amount of money.

Macros

Macros allow you to take a series of keystrokes and condense them into a single key press. For example, if you entered your name a lot, you could put your name in macro number 1. So let's say you tell your communications program to display macro number 1 by pressing the control key and the "1" key simultaneously (referred to as Control-1). When you press Control-1 your communications program would display your name and send it across the modem. Let's say you have the word "goodbye" saved in macro number 2; press Control-2 and the terminal program sends "goodbye" across the modem.

Macros can be a terrific time saver while on-line if you're a particularly slow typist. Even if you type hundreds of word per minute, macros can take some of the tedium out of modeming.

On-Line Timers

Many terminal programs come equipped with some pretty sophisticated stop watches that allow you to keep an eye on the amount of time you're spending on-line. These are used primarily when making a lot of long-distance phone calls to your favorite bulletin boards. The timers can usually be superimposed in a conspicuous place on your screen, so as you watch the seconds tick away, you realize your phone bill is growing larger and larger. Some of the really neato communications programs will even track how much time you're spending on various bulletin boards, so you can generate end-of-the-month reports. These reports don't serve any useful purpose, but they're great for hanging on your walls and impressing the heck out of your friends.

Host Mode

The host mode of a communications program turns your computer into a crude bulletin board system. While in host mode, the program basically sits around waiting for your phone to ring; when it does, the modem answers it and the program asks for a password. If the password is wrong, the user is disconnected. If the password is correct, the user has complete access to your system. Because the software works on a single password, there are no individual user accounts—so you have no idea who logged in. In most cases, host mode is used by PC owners with dedicated phone lines (phone lines just for their computers) so they can always access their important files—even from a remote location.

It isn't at all uncommon to see people leaving their personal computers on at work in host mode. Should an employee decide to work on a problem from home, he or she could dial into their desktop computer and download important documents or files.

Terminal Emulators

The term "terminal emulator" is synonymous with "communications program." The phrase "terminal emulator" was coined from the fact that your personal computer is mimicking (or *emulating*) a particular type of terminal. This analogy dates back to the Stone Age of communications, when mainframe computers would only talk to their particular brand of dumb terminal (for example, IBM mainframes would only communicate with IBM terminals).

The mainframe manufacturers were able to keep competing vendors from producing compatible terminals by programming their mainframes to use odd sequences of characters to perform special terminal functions—functions such as clearing the screen, "homing" the cursor, etc. Terminal emulators were written to allow desktop computers the ability to pretend they were a specific type of terminal.

For example, a DEC VT100 terminal emulator allows your personal computer to pretend it's a DEC VT100 terminal. As the sophistication of communications programs increased over the years, different terminal types were included in a single emulator package. This means today's personal computer can pretend that it is not only a DEC VT100 terminal, but an IBM and a TeleVideo terminal, as well. All you have to do is "tell" the communications program what type of terminal it should act like.

Although it's not critical that your communications program emulate a whole slew of different terminal types, it is a nice feature. Most of the more popular communications applications offer a nice selection of terminal types. The more, the merrier ....

Screen Buffers

Screen buffers are features in some of the nicer communications programs that allow you to browse information which may have already scrolled off your screen. Usually they hold from three to five screens' worth of data. At any time, you can suspend what you're doing and scroll *backward* to look at some of the old data. Once you become accustomed to being able to automatically recall information that has recently rolled off your screen, you'll wish DOS had a feature like this!

Screen buffers are a nice luxury, but don't let them be a major factor in selecting a communications program.

Chat Mode

As you become an old pro at calling local bulletin boards, you'll eventually engage in an electronic conversation with the system operator of a BBS. This conversation, known as chatting, takes place by both you and the sysop *typing* instead of talking. You simply take turns waiting for each other to say his or her piece.

Some programs feature what's called chat mode. This mode splits your screen in half—anything coming from the sysop is on the top half of your screen, and anything you type appears on the bottom half. This is helpful while electronically chatting because it makes it easier to see what each other typed; you won't confuse your words with the other person's.

If you chat with someone who is also using a terminal program with a chat mode, you can both type at the same time. Again, anything you type will be displayed on one portion of the screen, and anything your partner types will appear separately on another. This is neat because you don't have to sit around waiting for your partner to finish typing a sentence before you interject words of wisdom. If only verbal conversation worked this way.

Colors, Toggles and Other Neat Stuff

Most popular communications applications allow you to customize the program so it uses colors you find appealing—either because they're relaxing to your eyes, or because they just happen to look really funky. You can select almost any combination of foreground and background colors. If you want to be especially bizarre, select the color of the text and background to be the same.

The better communications programs allow you to toggle duplex, line feeds and other nifty things that may seem trivial at first, but which can be real lifesavers later in life. If you happen to make a small mistake while on-line, these features allow you to mask the problem so it isn't so apparent.

For example, when you call some bulletin boards for the first time, the system will ask if you require line feeds. If you say you don't need them when you really do, all the incoming text will be displayed on one single line. With a SuperMondo Communications Program, you can just have a few keystrokes tell your program to slip an extra line feed where it's needed. That will effectively mask the problem until you call the BBS again. By then you'll know the correct answer to the line feed question.

UPLOADS AND DOWNLOADS

In addition to sending raw information such as electronic messages or other miscellaneous text, your modem can also transfer entire executable programs along the phone lines. When you receive a program from another computer (such as an electronic bulletin board), you're *downloading*. When you send a program to a BBS or other computer, you're *uploading*. Both of these words will be used hundreds of times in any official gathering of modem freaks, so be sure you understand them thoroughly.

SOME RAP ABOUT PROGRAMS

Your executable programs stored on disk aren't much different from basic text files. From your disk's standpoint, programs are merely a collection of various characters held together in a very specific order. Keeping this in mind, if your modem can transmit a text file, why can't it transmit a program? (Don't answer that! It's a rhetorical question.) The answer is, "it can."

Programs can easily be sent or received to and from bulletin boards, or even a friend's computer, with very little effort. In fact, all you really have to do is answer a lot of silly questions for the communications software—and then sit back and watch *it* do all the work.

COMMUNICATIONS PROTOCOLS: X-MODEM

Programs are transmitted to and from other computers using what's called a communications protocol. A communications protocol is basically a strict set of rules two computers use when conversing with one another. These rules dictate when the sending computer is allowed to speak, and when the receiving computer is allowed to respond. The rules also specify how the two computers will speak to one another, by detailing the exact format in which the data is to be presented.

Communications protocols were primarily developed to allow for simple error checking. When modems first appeared on the market, the technology was new and, by today's standards, relatively unrefined. The early modems weren't at all tolerant of noisy phone lines and other bizarre conditions found in the real world.

Because modems translate noises into data, any stray background echoes or voices on the phone line will be interpreted as data. Modems can't differentiate between these stray noises and actual data tones being transmitted by another modem. As a result, you receive all sorts of extra characters on your screen. This unintelligible collection of garbage flowing across your screen is affectionately known as line noise.

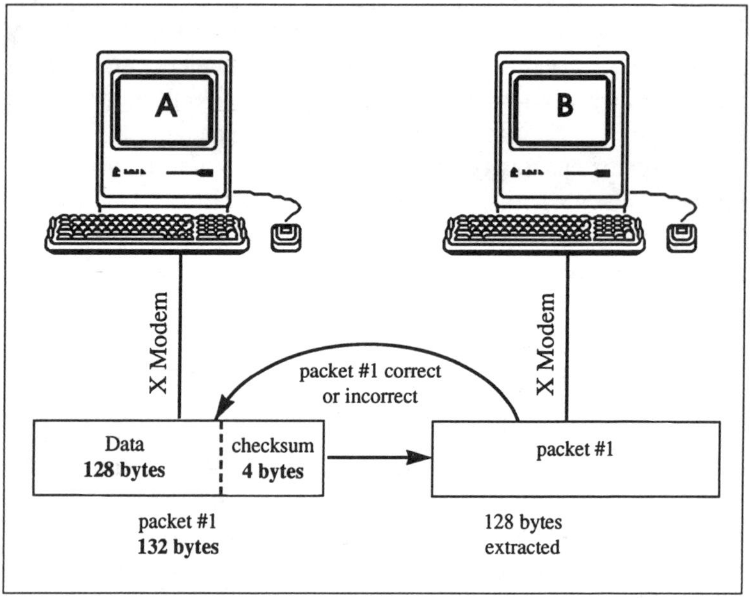

***Figure 7:** The X-Modem protocol isn't the easiest thing to demonstrate on paper, but this diagram should help. Computers A and B are both using the X-Modem protocol. This means that each computer is working with 128-byte blocks of data. In order to send and receive information there needs to be an additional 4 bytes added, the result is a packet of data. When computer A sends this packet of data, Computer B evaluates its integrity by performing a checksum and then either accepts or rejects it and requests another packet.*

If you're reading an electronic letter on a bulletin board and you see an occasional character (such as an open-brace or an underscore) in the body of the message due to a little line noise, you can still grasp the idea of the message. However, if any extra characters have snuck into a *program* file, the whole file is considered corrupted and it won't run. In most cases, your computer will lock up or crash if you try to run the program.

It wasn't at all uncommon to have to transfer a program three, four, or even more times just so you'd get a perfect, working copy. Considering that the fastest modems around during that era operated at 300 bps, exchanging programs via modems was a time-consuming and frustrating process.

To combat this line noise problem, Ward Christianson, a pioneer in the field of telecommunications, developed a communications protocol called X-Modem. X-Modem (sometimes referred to as Christianson Protocol) took the guesswork out of transferring programs. Using X-Modem, users could transfer their programs just once instead of three or more times; the protocol took all the mystery out of the process—you were almost always assured of receiving a perfect copy.

How Does It Work?

X-modem works by taking a program file and breaking it down in 128-byte blocks. A complex polynomial equation is applied to each block. The equation basically adds up the value of each character and then does some really bizarre mathematical stuff so the value derived is based solely on the characters in the block. After a process to figure out the results of the equation (called a checksum), X-Modem puts together what's termed a packet.

A packet is exactly 132 bytes long. It consists of the packet's number (such as 1, 2 or 3), the 128 bytes of data and the checksum (the result from the equation discussed earlier). This packet is transmitted to the receiving computer, where it's dissected.

The receiving computer first figures out the packet number, to make sure the two computers are still in sync. Then the 128 bytes of data are evaluated, and the *exact same* checksum formula applied by the sending computer is applied by the receiving computer. The receiving computer compares its results to the results sent. If the two match, the packet was received in perfect condition. If the two numbers don't match, the packet was received in error, and the receiving computer tells the sending computer to re-transmit the packet. Each packet is transmitted again and again until it's finally received without an error.

Varieties and Such

There are many different communications protocols used on bulletin boards, but almost all of them are variations on X-Modem. The most popular protocols in the BBS world today are X-Modem, Y-Modem, Z-Modem and Kermit.

Y-Modem is just like X-Modem, except it uses 1024-byte blocks instead of 128-byte blocks of data. This was done to reduce the amount of overhead needed for each transmission. In each X-Modem packet, we have 128 bytes of data and four bytes of overhead. With Y-Modem, you have 1024 bytes of data but still only four bytes of overhead. This means Y-Modem is spending more time transmitting data and less time transmitting such information as checksum values and packet numbers. Because overhead is so drastically reduced, Y-Modem is quite a bit faster than X-Modem.

Y-Modem also uses Batch Mode. Instead of having to transfer files one at a time, you can just give Y-Modem a long list of files and let it have at 'em. The protocol will transmit the names of each program (so they remain separate), as well as each program. Y-Modem Batch makes life a lot easier when you have quite a few long programs to send. You can simply start the transfer, and then find new and creative ways to kill time while the computer does all the slave work. Nothing could be easier.

Z-Modem is really bizarre, and it's not all that imperative that you know the nitty gritty involved in how the protocol works. For the record, though, Z-Modem is the fastest transfer protocol around—even a little faster than Y-Modem. Z-Modem isn't all that widely used right now, but it does promise to be the next big rage on BBSs nationwide.

Kermit, affectionately known as frog talk, is more than just a file transfer protocol. Kermit was developed and released into the public domain by the fine folks at Columbia University. Kermit is powerful, yet simple. Kermit can do relatively simple things such as transferring programs, or it can be used as a complex, yet easy to use terminal emulator. In terms of transmission speeds, Kermit is between Y-Modem and Z-Modem. Unfortunately, more depth on Kermit is well beyond the scope of this book, although there have been many fine books written on the subject.

CHAPTER 4

The Phone Line

As discussed in chapter two, a modem's purpose in life is to take data from your computer and pump it through an ordinary phone line where another modem/computer setup can receive and interpret it. While the modem is the most important single piece in the telecommunications puzzle, the phone line still plays a central role. Without a phone line, the modem would sit around with nothing to do.

How Much Does It Cost?

After you've invested in a computer and modem, usually the only ongoing expense is your phone bill. If you limit BBS usage to systems within your local calling area, your modem won't increase your phone bill one bit. This is because your modem can actually share the same phone line used for your household.

Your phone line treats your modem just like a voice caller. If you lived in sunny San Diego and instructed your modem to dial a bulletin board in Omaha, Nebraska, the phone company would bill you at the *exact* same rate as if you'd called your brother also living in Omaha. Ma Bell doesn't care if you're using your phone line to talk to another human or another computer—the charges are identical.

Along the same lines, while you're using your modem to communicate with an electronic bulletin board, you can't make or receive ordinary phone calls. So, if someone tries to call your home while you're using a BBS, the caller will hear a busy signal. If you suddenly decided to call someone while using a BBS, first you'd have to have your modem hang up the phone.

If you live alone, an additional phone line dedicated to your computer and modem probably isn't necessary for light BBS usage. However, if you share quarters with someone, I recommend paying extra for an additional line. It will make modeming much easier, and the convenience is well worth the additional cost.

TELENET: A MONEY SAVER

If you use a large commercial bulletin board, you will probably be able to save a significant amount on your long distance bill by subscribing to a service called TELENET.

TELENET allows you to dial a *local* phone number to electrically call another computer—even if it's located thousands of miles away. Here's how it works. The large commercial systems connect their phone lines to the TELENET service. You connect to TELENET by dialing a local phone number (TELENET has local numbers in almost every area code in the nation). Once connected to TELENET, you enter your account number, password and the name of the system you'd like to use. The TELENET communications system electronically channels your call to the specified computer.

No matter where the destination computer is located, you're charged the same hourly rate. TELENET's charges may seem high, but they're considerably less than if you dialed the other computer directly. Unfortunately, not every computer system in the nation is connected to TELENET; only the major commercial systems can be accessed this way. To save money while calling Mom and Pop bulletin boards, you'll have to utilize a different long-distance service, such as PC-Pursuit.

PC PURSUIT

PC Pursuit is a relatively new service offered by the same folks who bring you TELENET. PC Pursuit is considerably more affordable than TELENET, and it offers a lot more flexibility.

Basically, PC Pursuit gives you the ability to make unlimited long-distance phone calls for a flat rate of $25 per month. While this sounds great, there are a few restrictions. First of all, PC Pursuit is for modem calls only—you can't make voice phone calls using the service.

Second, the service is limited to major metropolitan areas. While you can dial *out* using PC Pursuit, your calls are limited to large cities such as San Francisco, San Diego or New York.

Finally, PC Pursuit is only available after hours. You have to wait until late evening before you're allowed to use the service. Actually, that isn't too much of a downfall, as most bulletin boards are busy during the afternoon and early evening hours anyway.

If you think calling bulletin boards all across the nation is glamorous, PC Pursuit is for you! It's a nice service if you want to do some long-distance bulletin boarding. However, most people find that local BBSs have just about everything they need.

CRIMINAL ACTIVITIES

One of the novice user's biggest fears about breaking into the exciting, dog-eat-dog world of telecommunications involves the phone line. The media does an excellent job of glamorizing nearly each and every story involving computer hackers who have devised a way around the phone company's security system, obtaining free phone service. Because of the media hype, a bit of paranoia has worked its way into the general public.

The most commonly exploited method used by hackers to get free long-distance phone service is through stolen telephone credit card numbers. Typically, a hacker will make a phone call and charge it to someone else's phone bill.

The telephone credit card numbers and access codes are very rarely stolen in a physical sense. The credit card numbers are easy to figure out. The account number is nothing more than the area code and phone number of the account you're charging against. A hacker can just randomly choose a phone number and he has an account number! Now he needs a secret access code that permits charges to be made against the account.

Don't let the phrase "secret access code" make you feel any more secure; the access code is simply a random, four-digit number. To charge a call to your personal phone bill, you dial a zero, the phone number you're calling, and then your credit card number and access code. Simple. In fact, it's the simplicity of the system the hackers exploit.

The theft of the credit card numbers and access codes usually follows this scenario: A devious programmer writes a program that attempts a "brute force attack" to figure out secret access codes. The codes are only four digits in length, so there are only 9,999 possible combinations. The hacker's special program tries to make a phone call and uses the access code "0001." If the phone company's computer says the access code is invalid, the hacker's program tries again using the code "0002" and "0003" if that doesn't work, then "0004," and so on ....

Once the hacker finds a code that works, he or she can use it for about a month. At the end of the month the unsuspecting person who owns the account is going to receive a whopping phone bill. After some calls to Ma Bell, the access code will assuredly be changed, so the hacker will have to search out a new victim.

The phone company will usually allow the innocent victim to slide on the unauthorized long-distance charges providing that evidence can be found proving they were fraudulently charged. However, stories like this are usually enough to frighten people away from modems, even though using a modem wouldn't make a person any more (or less) susceptible to such events.

PART THREE

CHAPTER 5

What Is a Bulletin Board?

An electronic bulletin board system (or BBS) is much like the old cork bulletin boards often found in supermarkets. The cork bulletin boards were installed so customers can post information in the form of index cards or flyers on the board. Other customers can glance at the board to see if there's anything of interest to them.

Computerized BBSs operate in exactly the same manner: Anyone can come along and electronically post information on almost any subject—from items for sale to requests for help on new software applications. However, BBSs offer a few advantages over their old cork counterparts.

For one, you can access an electronic bulletin board from the comfort of your own home or office. In fact, anywhere you have a suitably outfitted computer (one with a modem, communications software and a phone line), you can call and connect with a bulletin board. BBSs also offer you the ability to send *private* messages to other people. When composing a message, you can address the letter to all (meaning anyone who calls can see it) or you can tell the software to deliver the message to a specific user (that's called electronic mail).

Bulletin boards also offer you the ability to send (upload) and receive (download) software and other bits of information. Some bulletin boards have magazines or other pieces of literary genius available for viewing or downloading.

THE MAIN MAN

The person who runs an electronic bulletin board is called the system operator, a term often abbreviated to "sysop" (pronounced: *sisss-op*, rhymes with hisssss-op, if you pronounce it *size-op*, you're a dork).

Sysops are usually friendly folks, much like you and me. Instead of letting their computer sit idle when not in use, a system operator runs a bulletin board. Although some BBSs appear to be running on some kind of supercomputer, in reality they are mostly standard, run-of-the-mill personal computers. These host systems range from IBM 80386 systems to Amigas to Commodore 64s.

After you've been a part of the modem community awhile, it's pretty easy to spot the type of computer on which a BBS is running. The ultra-slow boards are probably running on a Commodore 64. The fairly quick systems are the Amigas, Atari STs and maybe the Apple IIgs. The ultra-fast bulletin boards are probably being run on either an IBM class of machine (like an AT or '386) or a Macintosh.

Nobody really knows what drives a system operator into running a bulletin board. It's almost always a "no win" situation; the sysop's computer and hard disk are subjected to a considerable amount of extra wear and tear (the computer has to be on all the time, and the hard drive is used every time someone calls). The sysop also has to install a dedicated phone line, and he or she loses quite a bit of access to the computer running the system. If a sysop gets a hankerin' to play *Flight Simulator* and someone is on-line, the sysop is going to have to wait until the user logs off.

Sysops also have the misfortune of being subjected to constant abuse. Instead of being thankful to a system operator for allowing use of the BBS (for free, even!), many users seem to feel that the sysop has the *obligation* to constantly entertain them.

SUMMARIZING A SYSTEM OPERATOR'S DUTIES

- **Verifies New Users.** This usually entails calling new applicants on the phone and making sure the person really exists. Some people like to play cute games and pretend to be other people.
- **Perform Regular Backups.** The extra wear and tear put on a BBS's host computer can greatly reduce the life span of a hard disk, so the sysop has to make sure all the system files, software, new messages, etc. are backed up. Considering that some BBSs boast having over 100 megabytes of disk space containing programs available for downloading, all that backing up can be quite a chore!
- **Policing the System.** The sysop is usually held responsible for the activities of his users. If a user posts a public message with illegal information (such as stolen credit card numbers), the authorities come after the sysop, not the user. So the sysop constantly has to watch for illegal messages, profanity, pirated programs and the like.
- **Improving the System.** Although not a requirement, sysops are usually programmers (if they don't start as programmers, they'll gradually pick it up through osmosis). The system operator is usually working on ways to improve his BBS to make it faster, more efficient, easier to use and more fun to call. This probably takes up more time than all the other duties combined.

While this certainly isn't a complete list of every task a system operator performs, it gives you a rough idea of how much time a person has to spend if he or she plans on running their own bulletin board. Other thankless duties include answering mail (usually from new users or people who forgot their password), fixing bugs in the software and paying the phone bill.

MESSAGE BASES

Unlike the cork bulletin boards of yesteryear, electronic BBSs are considerably more organized (and complicated). Cork bulletin boards have messages on a wide assortment of topics sitting in a single, common area. BBSs, on the other hand, ordinarily have sectioned-off areas where messages on the same general topic are kept together.

For example, when you tell a BBS you want to read public messages, you'll have to specify on which area you want to read. Imagine each area as a separate cork bulletin board. On top of each of these separate bulletin boards is a sign that indicates what kind of messages can be posted (such as "for sale" or "wanted"). Apply this analogy to BBSs and you have what's called a message base—each of these areas is called a message base.

There are usually a multitude of message bases on any BBS. The topics range from "general chit-chat" to "programming topics" to "for sale" to "tips and tricks for *Flight Simulator*." If you frequent a bulletin board that doesn't have a message base set up for a topic you think would be of interest to others, just tell the system operator. Most sysops are more than happy to hear (and act upon) suggestions from users.

ELECTRONIC MAIL

In addition to being able to compose public messages on bulletin boards, you can also send private electronic letters to other BBS users.

Electronic mail is similar in many ways to the United States mail in that you send and receive letters. However, electronic bulletin boards deliver your masterpiece immediately! As soon as you finish typing the letter on-line, the BBS usually delivers it. The next time the addressee calls the BBS—whether in five minutes or in five days—the electronic document will be sitting there waiting.

Some bulletin boards offer extras in their mail system, such as certified mail. When you send an electronic letter via certified mail, the system will *automatically* send you a letter citing the date and time your original letter was read by the addressee.

Other neat features found on some of the better BBSs include:

- **Message Forwarding.** This feature gives you the ability to forward a letter to another user (or group of users) after you've read it.
- **Carbon Copies.** As the name implies, this allows you to send copies of your letter to more than just one person.
- **Blind Carbon Copies.** Blind carbons, as they're sometimes called, are similar to regular carbon copies—except only the person who receives the "blind" carbon copy knows additional copies were sent. For example, if I wrote a letter addressed to you and I listed Fred Flintstone on the blind carbon copy list, both you and Fred would receive the same letter. But *you* wouldn't know that Fred received a copy—it wouldn't be in the message's header anywhere. Fred, on the other hand, would see from the message's header that the letter was addressed to you and he received only a blind carbon copy. Blind carbons may sound extraneous—but it's a nice feature.
- **Distribution Lists (or Aliases).** Some of the super-slick BBSs let users create their own *distribution lists*. A distribution list is basically just an alias for a group of users. For example, I tell the BBS that if I address a letter to "writers," I really want the letter to be addressed to Jim Kimble, e.e. cummings, Tom Clancy and Steven King. It's a lot easier to just type the word "writers" versus having to type (much less remember) all the other names.

- **Networked Electronic Mail.** If you happen to come across a BBS that boasts being networked with other bulletin boards, you'll probably have access to network mail. Networked Electronic Mail works exactly like ordinary electronic mail, but with one exception: the receiving user doesn't have to use the same BBS you use! You simply tell the mailer what system your friend uses and your friend's name. The mail system will forward the letter to their BBS where it will be delivered. The disadvantage to networked mail is you must already know the name of the user you're sending to; you can't find out someone's name on another system without having to call it directly. Networked mail is something that's really catching on, but it's not something you run across every day.

ACCOUNTS AND PASSWORDS

Because electronic BBSs offer features such as private electronic mail, passwords are needed so the bulletin board can confirm users' identities. Without passwords, *anyone* could connect to a bulletin board and claim to be, say, Jim Kimble, and then read private mail addressed to me.

Most BBSs allow you to choose a password when you register with the system. I strongly recommend using a different password for each system you call. Although it's very uncommon, a bulletin board's password file can be broken into; if you used the same password on every system, an unscrupulous person might be able to login to BBSs all across town saying they're you!

Some bulletin boards will choose your password for you by picking an eight- or nine-character string of random numbers and letters. While this ensures your password will be unique, the computer-generated passwords are usually pretty tough to remember. With the advent of the automatic login scripts (see Chapter Three), you can program your communications program to enter your name and password automatically.

Login scripts are great if you're the only one who uses your computer. If other folks (like co-workers) dabble around on your machine, you might want to think twice about keeping login scripts on your disk. All the person at the keyboard would have to do is look at the login script and he or she would know your passwords to whatever BBSs for which you happen to have written scripts.

I use login scripts on my machines at home because I'm the only one who uses them. However, while my computer at work may be "mine," someone could easily tinker around with it while I was goofing around at the water cooler. Because of this, I manually enter my account number and password when calling a BBS from the office. That means I have to remember each and every password for each and every bulletin board I call. Bummer.

UPLOADS AND DOWNLOADS (FREE SOFTWARE)

Bulletin Boards give you the ability to receive (download) programs to your home computers; they also allow you to send (upload) programs to the BBS.

Different bulletin boards have different policies on downloading software. Some systems have huge vaults of programs and users are allowed to grab anything their little hearts desire. Other BBSs, however, limit downloads to a specific number of programs or bytes (like 100,000 bytes) each day. Finally, some BBSs require users to upload programs before they'll be allowed to download any; usually at a ratio of something like 1:2—one upload "buys" two downloads.

Sysops have to implement policies to police users downloading software to prevent what's called a "leech." Leeches are those disgusting users who call all sorts of bulletin boards and do nothing but download the latest and greatest software. Most system operators don't care if you download programs—that's why they're there—but they usually ask that you repay the board by either uploading some new programs you have, or by contributing to the discussions in the various message bases.

CHAPTER 6

Calling Your First BBS

You can find the phone numbers to local bulletin boards by either referring to local computer magazines (such as San Diego's *ComputorEdge*), by asking the dealer where you purchased your modem, or by attending a local users' group meeting.

In the front of your White Pages you'll find a table of phone number prefixes that are local (read: no toll charges) to you. If you don't pay attention to the phone numbers you're calling, you'll probably receive a big surprise at the end of the month when your phone bill arrives.

Once you've decided what BBS to call, go ahead and have your modem start dialing. Some communications programs will do that for you. If not, you'll have to manually instruct the modem to start dialing with a statement like "ATDT555-1212".

Provided the bulletin board you're calling isn't busy (meaning someone else is using the system), you'll receive some kind of indication from either the modem or communications software telling you that you have successfully connected with another computer. Usually you'll see a message like "CONNECT 2400," modem talk for "connection established at 2400 bps."

LOGGING IN

After connecting, you may have to press your return key a few times to wake the BBS up. That's because a BBS program sits idle waiting for someone to call. When you press the return key, the BBS software is told to start doing its thing.

The first thing the bulletin board should do is display a welcome banner that contains information such as the name of the system and perhaps some simple login instructions. Some BBSs will ask you for your name, while others will ask for an account number. Most of the systems will say something to the effect of "New users enter 'NEW'." If you're a new user (and you are), type the word "NEW" (don't include the quotes) and press return. If that doesn't spark any kind of reply, try the word "GUEST."

```
2400
                       Welcome to ComputorEdge Online!

ComputorEdge (yes, we really spell it that way) is San Diego's computer
magazine. We are published on alternate Fridays every month except
November and December. This system is updated sometime after the issue
comes out on Friday.

Some commonly needed mail addresses are: TINA BERKE and WALLY WANG
(ComputorEdge Editors) and SKOSCH PENROSE (On-Line Editor)

Your sub-sysops are: KENNETH BIBB (UNIX), JOE CROSS (Amiga), AYMAN IBRAHIM,
(Apple).

First Name?
ALT-F10 HELP | VT-52 | FDX | 2400 N81 | LOG CLOSED | PRT OFF | CR | CR
```

Figure 8: The above is a good example of the first thing you will see after dialing the number for an on-line system or BBS

After you enter the magical key word, the bulletin board will start a registration process. Here you'll be asked for information such as your name, the type of computer you're using, any special characteristics your computer may require (added line feeds, nulls, etc.), and the password you'd like to use.

The system operator ordinarily requires you to release your phone number so he or she can call and verbally verify that you're you. This is an attempt to avoid misrepresentation (you'd be amazed at the number of people who call new bulletin boards claiming to be Jim Kimble. Ahhh, kids ...).

Immediately after registering, you're usually allowed to start using the bulletin board—although you'll probably have limited access until the sysop has a chance to validate your registration data.

CHOOSING A USERNAME

Some bulletin boards don't refer to users by their real names, but by pseudonyms (like aliases). Usernames are a lot like handles on the old CB radios; in fact, that's probably where the idea started.

In my opinion, usernames give punks and troublemakers a way to hide behind a cloud of anonymity. Some BBSs require usernames that make sense; one BBS I encountered was oriented toward HAM radio operators, so everyone's username was their call-sign. Other systems have stupid usernames, particularly the BBSs that cater to the software pirates of the world.

Anytime you run across a user whose first name is "The," you know you've found a loser. You'd be amazed at some of these usernames: "The Disk Doctor," "The Black Knight Of Death" and "The Pirate," are typical. I always use my real name when calling a BBS, and I usually try to avoid the systems that require or encourage others to use usernames.

ADMINISTRIVIA

After registering (applying for access), you'll go through what I call the Administrivia Process. That's where you'll be presented with all sorts of neat "new user information." You will probably see a few bulletins containing information such as the rules for the BBS (i.e., no profanity, no messages with illegal content, etc.), hours of operation (if not 24 hours), policies regarding uploads/downloads and general announcements.

```
Searching Message Base ...
You have no personal messages waiting.

ComputorEdge On-Line Main Menu
==============================
<R>egister Your Account
<E>lectronic Magazine
<U>ser Utilities
<G>oodbye

Command: r
User Registration
=================

You must answer the following questions correctly to
receive full access to this system.  Thank you.

Your Full Name: Scott Millard
Daytime phone number: 555-3352
Enter your street address and number:
ALT-F10  HELP | VT-52 | FDX | 2400 N81 | LOG CLOSED | PRT OFF | CR | CR
```

***Figure 9:** This figure shows the process you must complete in order to become a known user. This is the first time Scott has called the* ComputorEdge On-Line *system.*

Somewhere in there you'll probably see a friendly request for donations. From reading the previous chapter, you know that almost all the public bulletin boards are run by hobbyists like you and me. It's very rare to run across a system with some kind of corporate sponsorship.

Because most BBSs are run merely as a hobby, the sysop is left with quite a few bills at the end of the month. To help pay for these expenses (electricity, additional phone line[s], etc.), system operators sometimes ask that you make a one-time donation of $5 or $10.

These donations are usually voluntary, so you're generally not looked down upon for failing to make a monetary contribution.

However, sysops often try to make donations more appealing by offering such goodies as extra time on-line, or a higher level of access to things.

THE MAIN MENU

Most of the world's bulletin boards are menu-driven. This means every option available to you is presented on a menu. You instruct the BBS what to do by pressing a letter that corresponds to a specific command. A sample main menu might look something like this:

If I wanted to read any announcements the system operator may have written, I would press the "A" key and hit return. Thc BBS software would display the new announcements.

If I wanted to have the system check to see if I had any personal messages (electronic mail) waiting, I would press the "R" key. If I did happen to have mail waiting, the system would display a submenu that contained a whole new list of options. For example:

Read messages:

<F>orward
<R>everse
<N>ew messages
<I>ndividual messages
<Q>uit to main menu

Looking at the submenu, I can elect to read my messages forward (meaning I can read them in the order in which they were written), reverse or individually. If I selected "new messages," the BBS would display only those mail messages I haven't already seen. Generally it's a good idea to delete your mail messages after you've read them, otherwise they needlessly take up valuable disk space. Here's a summary of some popular commands:

<A>nnouncements
<C>hat with the sysops
<G>oodbye (hang up)
<M>essage bases
<R>ead private mail
<S>end private mail

Chatting

Nearly every bulletin board you call will have an option titled "chat" or "chat with sysop." Chatting on the BBS is like chatting in real life—two people talk to each other. On a bulletin board, the caller (that's you) and the system operator (if he's around) talk by using keyboards instead of mouths.

You type a question, comment or whatever, and the sysop replies. Like a telephone conversation, only one of you can talk (type) at a time.

A piece of friendly advice: page the system operator only when you have a real question/comment. Sysops *hate* users who page them just to say "hi!" If you have a simple question or comment that doesn't need to be answered immediately, you'd probably be better off sending mail instead of chatting, especially if you're a slow typist.

G-Files (Text Files)

The term G-Files comes from the Commodore 64 BBS software called C-Net. However, some of the other popular bulletin board programs have adopted the nomenclature.

G-Files are simply text files. When you tell the system you want to see the G-Files (or text files), you'll be given another menu that lists each file available. These files typically contain helpful bits of information—everything from tutorials on assembly language programming to listings of every bulletin board within 183 nautical miles.

Message Bases

Messages bases are the BBS's public message area. After telling the system you want to read the message base(s), you'll be asked *what* base you want. Typically the message bases are sectioned off by category.

The system should give you a list of each available message base (if not, try entering a question mark). From the list you can decide on the base that seems most interesting to you. The message bases usually have topics like general chit-chat, political discussions or perhaps programming tips.

Be sure any messages you post to a particular message base follow the base's subject. Otherwise, you're asking for a mailbox full of hate mail.

Voting Booths

Some BBSs have cute little voting booths that allow users to voice their opinions on various issues. Usually the voting is concentrated on issues concerning the bulletin board. For instance, some sysops want the users' input before adding or removing a feature.

Other Commands

The menu systems make bulletin boards incredibly easy to use. The more complicated a BBS, the more on-line help you'll find offered. If you ever wonder what a particular command does, just try it!

SOME RAP ABOUT BBSs

Each individual bulletin board is a lot like a miniature, self-enclosed city. And, as with cities, every bulletin board is different. Many of them may look and feel similar, but sysops are almost always adding little goodies to customize their software.

Because of the vast differences between systems, it's nearly impossible to write a step-by-step tutorial for every BBS. You'll have to look around and play with each of the commands on a new BBS to figure out what they do and how they work. It may seem like a lot of work, but actually that's half the fun of calling a new bulletin board.

BBS Acount Information

Be sure to write down your BBS account number and password in a safe place. After being validated, you'll be able to enter merely your account number (or name) and your password. The BBS saves your terminal settings (settings for how many columns your screen displays, what type of computer you're using, and such) so you don't have to enter this information every time you call.

CHAPTER 7

Commercial BBSs

A long, long time ago in a city far, far away, a modem user logged into a public electronic bulletin board and was suddenly hit with a brilliant idea. "What if I set up a really neato BBS of my own and charged people for access?" The geek thought to himself, "I'll make millions!"

Sure enough, the geek invested in some state of the art hardware and diligently set out to write *The Perfect BBS* program. After a few months of rounding up investors to share the cost of the hardware, and after some intense programming, the first commercial BBS was born.

Since that fateful day many years ago a flurry of commercial BBSs have been popping up all across the nation. Every now and then one of them finds its niche in the BBS world and actually survives—maybe even flourishes. But for every one that lives, many more fall by the wayside.

The main reason these systems don't succeed is that they're really not needed. Nearly anything you can get from the big commercial systems, you can get from a public (read: free) BBS right in your own town. And for those little knick-knacks you can't obtain locally (such as traveler's reports, up-to-date weather information, on-line shopping, airline ticket information, etc.), you can look to one of the already-established and less expensive commercial services. It's just too difficult for a new system to break into the already over-populated market.

HOW DO THEY WORK?

Commercial bulletin boards operate just like normal BBSs, except these systems charge a usage fee. Almost all major systems (as I'll discuss later in the chapter) structure their billing system like the phone company's. You're billed by how many minutes you use, and are given certain discounts for using the system during holidays and non-peak hours. Nearly all the systems charge a monthly minimum of around $15 to $20.

Some of the newer systems feature a different billing method. Prodigy, for example, bills its users a flat monthly rate. You don't pay for connect time (time spent on the host computer), so your monthly bills are the same amount each time. It doesn't matter if you didn't use the system at all during the month, or if you were on-line 18 hours a day; you pay the same low fee (usually only about $10 per month). This billing method is a lot nicer compared to the minute-by-minute rate structure, because you aren't sitting at your keyboard feeling rushed, especially if you're a slow typist.

WHAT DO THEY OFFER?

The major commercial systems offer quite a few features not found on local BBS systems. You'll get a *huge* readership for your hard-to-answer questions, for one. Some of the systems have tens of thousands of users, so if you post a technical question to a popular message base, you can expect to see a multitude of responses.

Some of the most attractive features include timely weather reports, sports scores, stock market information, on-line shopping services and airline ticket reservations (and even purchasing, if you charge the ticket to your credit card).

Each commercial BBS offers recreational activities as well,

such as games and electronic chatting (discussed in more detail later). Plus you get hundreds of megabytes worth of software available for downloading. However, here's where the big services nail you to the wall on your bill: if you're paying on a minute-by-minute basis and you decide to download a 300,000 byte file, you're going to spend 30 minutes to an hour waiting for the file to transfer—and that file could end up costing you a pretty penny!

THE PHONE BILL

These large systems are usually located somewhere in the central United States, so you may be wondering how much additional money you'd end up spending in telephone bills. Lucky for you, this has been taken care of ... sorta.

Many of the really large services provide a whole load of phone numbers within each area code across the nation. Simply dial this local number, and through the miracle of electronic switching, you're automatically channeled to the main computer. Best of all, you pay only for the local phone call.

If the BBS in which you're interested doesn't provide this feature or just doesn't have a number in your area code, you'll have to use a service called TELENET (see Chapter Four, The Phone Line). To use TELENET, you dial a local phone number, tell the system who you are by entering an account number and password, and then connect to the computer service of your choice regardless of its location.

The beauty of this is that Ma Bell only charges you for the *local* phone call. TELENET will send you a bill at the end of the month for the charges incurred, but TELENET's rates are considerably cheaper than regular long-distance rates.

HOBBYIST TURNED ENTREPRENEUR

In the early days of BBSs, it was absolutely unheard of for a system operator to charge his users for access to a bulletin board.

However, most of the users understood how expensive it is to run a board, so they often sent the sysop unsolicited donations.

Some greedy folks decided that their BBS, started as a hobby, could perhaps be the source of a financial fortune. Usually these sysops first require you to pay a one-time "donation" for access, and then slowly move on to charging a monthly fee or charging for connect time.

Considering the fact that most major cities have well over 100 local BBSs (San Diego County has at least three times that), a bulletin board would have to be incredibly popular to draw users willing to pay for access. Face it, who would pay to call a local BBS when there are 99 others just like it in town—but free!

For this reason, it's not really common to come across a BBS that charges money. In the San Diego area, all but a couple Macintosh bulletin boards have some sort of monthly fee (about $10 or $15). Users who elect to pay the fee are given access to the well-stocked download area; people who don't donate are still allowed to call the system, they just can't get any software.

COMPUSERVE

Of all the large commercial services available, CompuServe is probably the best. The hourly rates are pretty reasonable if you call during the non-peak hours (from 6 p.m. to 6 a.m., weekends and holidays), and the service is accessible via TELENET.

CompuServe offers the typical assortment of neato features expected with commercial BBSs, but it also has a well-stocked On-Line Mall. From the comfort of your home, you can purchase anything from a pair of socks to an automobile. Your orders are automatically charged to your credit card.

CompuServe is notorious for its CB or on-line chatting, organized similar to the old CB radios so popular a few years ago. You simply pick a handle by which you want to be known, and then select one of 40 channels.

While in the CB area of the system, you can electronically chat with other users. You can talk about anything from the bugs in *WordPerfect* 5.0 to the weather in Johannesburg. It's neat to be sitting at your computer communicating with people in locations ranging from Crab Apple Cove, Maine to San Diego, California.

CB is neat, but be careful while using it! You'll sit down for a brief chat with someone and the next thing you know, you'll realize you've been typing for three hours, and at $6 or more per hour on CompuServe, that can get mighty expensive!

Subscribing to CompuServe is a piece of cake. You can head to your favorite computer book store (or computer dealer) and buy a sign-up kit for the service for about $35. The kit includes all the information (local phone numbers, an account, a password, etc.) to get on-line. The $35 you spend for the kit is credited to your account, so you get $35 worth of free on-line time!

THE SOURCE

The Source is the largest of the commercial BBS services, the self-proclaimed best in the industry. Personally, I feel The Source is over-priced and over-rated. Everything offered on The Source can best be obtained from another service, at a significantly lower price.

The Source has the highest per-minute rates of any service, and the monthly minimum is considerably higher than most of the other systems. I advise against subscribing to The Source.

GENIE

GENIE is nothing more than a rip-off of all the other services that already exist. Yet I've heard nothing but good about GENIE from those people who are currently using the Genealogy Forum. This forum allows you to discuss such genealogy issues as whether any other users might have leads on certain family trees.

If you're heavy into genealogy, this is something you'll want to look into! Imagine being able to ask the help of thousands of people located all across the world—and from the comfort and convenience of your desktop.

BIX

BIX has had all sorts of trouble in its shaky past. Other than users who are allowed to use the system for free because they moderate conferences, not all that many people call BIX.

The service has a really neat conference system, but that's its only real selling point. I've never been that impressed with BIX.

MCI MAIL

MCI Mail really isn't a bulletin board, but merely a convenient way of sending electronic documents to other users worldwide. CompuServe and MCI recently signed an agreement whereby users on one service can send letters to users on the other service (i.e., you can send a letter from CompuServe to your friend who uses MCI Mail, or visa versa).

MCI charges by the quantity and size of your letters. Most of the people using MCI Mail are business folks communicating with other business folks. Every now and then you'll see someone who puts their MCI Mail address on their business card.

MCI Mail really isn't the type of service that would interest the typical hobbyist. However, if you have a need to communicate with a lot of different people who are existing users of the service, this might be something to consider. Lots of independent consultants and freelance folks (such as writers, programmers, etc.) use MCI Mail to keep in touch with clients and agents.

PRODIGY

Prodigy is a relatively new service offered through a joint partnership between Sears and IBM. Prodigy is one of the few systems charging by the flat-rate monthly fee (around $10 per month). Prodigy makes its money by selling on-line advertisements. Users are constantly shown half-screen adds for all sorts of products.

Prodigy is primarily for on-line ordering. You can order any product sold by Sears and have it charged to your credit card, just like ordering out of a catalog. However, other features keep users entertained (and thus calling back): on-line games, electronic mail, movie reviews, etc.

The neat thing about Prodigy is its graphics. When you sign up for Prodigy, you get this software program that takes control and acts like a funky graphics interpreter. All the screens you receive from Prodigy are full-screen graphic designs. The graphics are crude, but they're fun to look at.

SOME RAP ABOUT COMMERCIAL BBSs

Personally, I don't like the commercial bulletin board systems because I think they're a rip-off. While I concede that these systems offer a few things not found around town (like up-to-date weather reports, airline ticketing information, stock market quotes, etc.), the cost you pay for these conveniences can be quite excessive.

If you wanted to try one of these services just to see how neat they are, try Prodigy. It's cheap, it's user friendly and it has many of the features offered by all the other services—plus some they don't have.

My advice to anyone who tries one of the systems that change on a minute-by-minute basis: Be careful! The commercial services (especially the huge download libraries) can be very addicting, and it's not hard to run up a bill of $100 or more.

CHAPTER 8

Starting Your Own BBS

After you start playing with electronic bulletin boards on a regular basis, you may find yourself asking "Why don't I start a BBS of my own?"

Just about every new modem user goes through a phase where he or she seriously considers starting his or her own bulletin board. My advice? Ignore your feelings; as with every other phase you've gone through in life, this too will pass.

As any experienced system operator will tell you, running a BBS requires an incredible investment of not only time, but money. You don't have to be an expert programmer to be a good sysop, but a little bit of programming knowledge will make life a lot easier. System operators are constantly being plagued with requests from users, and many of the requests require custom enhancements to the BBS software—enhancements that can only be accomplished by some serious hacking to the BBS's programs.

If every aspiring sysop in the world were to follow my advice, there would be no new bulletin boards. So for those brave few who feel determined to make the On-Line Community a better place by starting their own BBS, read on ....

HARDWARE INVESTMENT

Having the necessary hardware is probably the first (and easiest) requirement to meet when setting up a new bulletin board. Basically, you'll need the following major items: a computer with a decent-sized hard drive (I'd recommend nothing less than 20MB, although 40MB or more is preferable), a modem and a phone line.

Hard Drive

While the modem is certainly important, I'd say the disk drive is probably the most significant piece of hardware. Most people's personal hard drives are small (small, in bulletin board terms, means less than 40 megabytes) and purchasing a larger drive represents a considerable monetary investment. But a large drive isn't a *luxury* in running a BBS, it's a *necessity*! The more disk space you have, the more messages your system can store; the more electronic messages your BBS can handle; the more users you can accommodate; the more programs you can make available for downloading. A large hard drive is important.

Some seasoned system operators out there will probably tell you how they run their city's most popular BBS on nothing more than a slow 10MB drive. Ignore these people—they're the exception rather than the rule. Any new BBS is going to have a hard time getting started; there's a lot of tough competition for new users among fledgling bulletin boards. Having a solid collection of fine-tuned hardware may be just what you need to draw users. It could make the difference between being the hottest BBS in town or being just another "here today, gone tomorrow" bulletin board.

Modem

The next piece of hardware you may want to upgrade is your modem. Let's face it, modems are cheap and easy to come by; for that reason you should seriously consider investing in a 2400 bps monster if you don't already have one. A lot of users don't take

BBSs running at 1200 bps seriously (not to mention 300 bps systems).

The faster your modem, the less time users are going to be spending on-line. That means you can accommodate more users each day! If you plan to run a BBS boasting a huge download section, a 2400 bps modem is a necessity. Your users will go crazy waiting for a huge program to transfer at 1200 bps!

The Computer

The final piece of hardware needed to run a BBS is your computer. Almost every computer in the industry has a BBS program written for it. In fact, much BBS software is either public domain or shareware. The faster your computer, the better. I'm not going to discourage anyone who owns what might be deemed a "lesser" computer brand against starting a BBS, because a good bulletin board can be run on any kind of system. The users make a BBS, not the hardware. Case in point: San Diego has a very popular BBS being run on a Timex Sinclair computer!

To sum it all up, the faster your computer, the better. Users hate waiting around during slow response times. The faster computers (such as IBM clones and Macintoshes) can usually do neat tricks the slower beasts can't, such as running BBS software that allows for two or more callers at the same time!

SOFTWARE INVESTMENT

After you've put together your collection of hardware, you still have a tough decision ahead of you: What type of bulletin board *software* are you going to run? Many different BBS programs are available for nearly every computer in existence, and each has its advantages and disadvantages.

The software *is* the BBS, so you must investigate the strengths and weaknesses of each available bulletin board program carefully. Some users will avoid a certain BBS program like the plague,

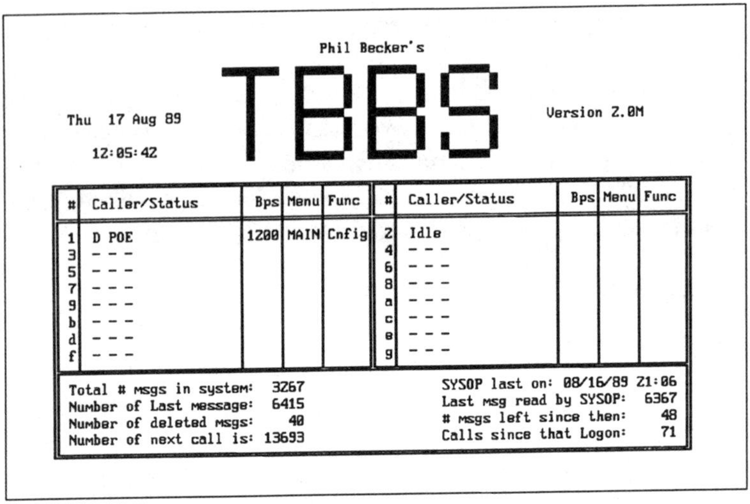

Figure 10: *For those of you wondering what BBS software is really like, here is an example of the status screen on the* The Bread Board System, *a bulletin board software package. Notice that there are two phone lines supported; the left side has a caller and the right side is idle.*

while others will be drawn toward *your* system simply because you're using a software package with which they're familiar.

An in-depth review of each popular bulletin board program available for even only *one* computer system would be well beyond the scope of this book. Instead of turning to a book for reference, I recommend calling all sorts of different BBSs in your area and finding out what *you* like and don't like about each particular system. After you've investigated the various packages, you'll be able to make an informed decision.

PERSONAL INVESTMENT (TIME AND MONEY)

Your new BBS is going to drain many, many hours of free time out of your already-busy schedule. Running a BBS means having to perform such tasks as verifying new users, performing system maintenance and periodic backups, trouble-shooting defective hardware, and logging in at least once a day to answer routine messages.

The sysop must work hard to keep the system secure, to keep the BBS clean of profanity and offensive material, and to keep users happy. This is an ongoing job that requires many hours of time each week; the bigger the BBS, the bigger the job of the sysop.

The sysop, being the sole proprietor of a BBS, is also responsible for *all* the bills. You'll have a monthly phone bill to which you can look forward, as well as the expense of having to replace any hardware (such as disk drives, computer components or modems) that may fail from routine wear and tear.

A PHONE LINE

If you're going to start a new BBS, plan on getting a dedicated phone line for your computer. Trust me. Sharing your voice line just doesn't work. The extra phone line will probably only cost you about $25 per month, provided you don't make any long-distance phone calls with your computer.

A GOOD GIMMICK

Some cities in the nation have experienced a bulletin board population explosion. San Diego County, for example, has well over 250 BBSs in operation at any given time. With such a large base of systems already in existence, it's hard to imagine how a fledgling BBS could offer anything new to users. For this reason, a new or aspiring sysop must come up with some kind of gimmick that makes his bulletin board better.

Because the interests of thousands of modem enthusiasts located across the nation are so diversified, it's impossible to design a BBS that caters to everyone's personal needs and desires. Instead of trying to please everyone, you should concern yourself more with finding a niche and specializing your bulletin board.

Some BBSs focus on aviation, some on cooking, some on gardening, some on real estate—the possibilities are endless.

Although you should focus your bulletin board toward specific topics, don't forget to keep at least one general message base where users can post messages about anything. As long as you have a general message base, you will still draw the vagrant users who are looking for a "home" system.

Advertising

Once you've organized your BBS and tested the heck out of it, it's time to start advertising your phone number. The best way to go about this is to call some of the popular systems around town and make your number known. Some BBSs have on-line lists of your city's bulletin boards; submit your board's name and phone number so you can get on that list!

Users groups often publish lists of local BBSs in their newsletters. You might want to explore that avenue.

GETTING DOWN ON DOWNLOADS

Supporting a healthy download section on your BBS is an excellent way of luring new users to call your system. However, users who call a system strictly because of downloads are often more of a problem than a help.

Leech

The BBS community has a breed of user called a "leech." Leeches like to call bulletin boards and download all sorts of software without ever reciprocating by uploading new software or contributing to the discussions on the message bases.

While leeches are busily robbing your pantry of software, your other users can't connect to the BBS. Because downloads (by both leeches and normal users) eat up so much time, some system operators disable downloads for slow modems. Quite a few sysops don't allow users calling in at 300 bps to download software; some don't let 300 bps callers login at all!

RUNNING A BBS

As this chapter has hopefully illustrated, running a BBS isn't an easy task—it takes a lot of time, money and patience. Nearly every modem user has given serious thought to starting his or her own bulletin board, so you're not alone. But before you do anything about your dream, be sure to sit down and take a long look at what's involved. Call your favorite BBS and ask the sysop some questions: How much time does it take out of his schedulc? How much are his phone bills? What kind of problems does he face?

You might be surprised at the answers!

PART FOUR

CHAPTER 9

Often-Asked Questions—Answered!

The preceding chapters have explained the fascinating world of modems and modeming. Now you know enough about telecommunications to start asking some questions. Below are questions (and answers) that are most commonly asked and answered. Hopefully yours will be in here too.

WHAT DO FILE EXTENSIONS SUCH AS .ARC MEAN?

When you're perusing the software library on bulletin boards, you'll usually see file names with the extension .ARC instead of .EXE or .COM. ARC is short for ARChive. By running a special program, you can actually compress (or archive) a file down in size—sometimes by more than 80 percent! The smaller the file, the less time it takes to download, and the less space it takes up.

The file compression utilities that perform the file archiving and un-archiving perform other functions. For instance, some software packages (such as the Norton Utilities) are made up of a whole slew of small programs. Instead of forcing you to download each and every one of the small programs, an archive program will take each of them and join them together to make one big program, and then compress the big program. When you un-archive the big program, it will be returned to its original size, and then the smaller programs inside it will be broken off. When the un-archiving process completes, you have your original small programs.

The packing and unpacking programs are available on most BBSs in regular (.EXE and .COM) format.

HOW DO I "LOGOFF" A BULLETIN BOARD?

Most BBSs use one of three methods. The most common method is to select the "G" (Goodbye) option from the main menu. If "Goodbye" isn't an option, try "O" (for Off). Failing that, leave the sysop a letter asking him how to logoff and hang up.

SHOULD I SET MY MODEM TO ORIGINATE OR ANSWER MODE?

This should be taken care of by your communications software. However, if it's a problem, the modem that originates a phone call should be in originate mode—and the modem that answers the phone should be in answer mode. That's the simplest way to remember.

WHAT'S A C-NET?

C-NET is a type of bulletin board that runs on the Commodore brand of computer (i.e., Commodore-64, Commodore-128, etc.). The C-NET BBSs are rather slow. That's not to say they're all bad, but most of them are. They're a good beginner's bulletin board, but you'll outgrow the systems eventually.

WHAT BRAND OF MODEM IS BEST?

As long as you stick to a name brand such as Hayes, U.S. Robotics, Micom, MicroComm, etc., you'll be all right. The cheapie modems work, but they don't offer you anywhere near the functionality of the others.

SHOULD I WORRY ABOUT VIRUSES AND TROJAN HORSE PROGRAMS?

Yes! Viruses and Trojan Horses are spread primarily by way of electronic bulletin boards. The best way to protect yourself is to test software only after first running a disk-protection program such as NOHARD or HARDOFF. These programs tell your computer to ignore the hard drive and not allow a program to read or write to it. You can test new programs without using your hard drive. If, after careful examination, everything looks good, you can simply reboot your computer to regain full hard disk access.

WHAT DOES IT MEAN WHEN SOMEONE PUTS THE CHARACTERS :-) IN THEIR ELECTRONIC MESSAGE?

When you say something sarcastic, other people in the room can usually pick up on the sarcasm by the way your voice fluctuates. Unfortunately, with electronic messages, other people can't hear your voice, so they don't always know if you're serious or just kidding. The text :-) is supposed to be a happy face on its side (turn your head to the left), and means "just kidding" or "sarcasm alert."

A lot of variations on the original happy face have been introduced into the BBS community over the years. They all basically mean the same thing.

WHAT'S A SYSOP (SYSTEM OPERATOR)?

A system operator is the guy or gal who operates a bulletin board system—the person who runs the BBS! You pronounce "sysop" sisssssss-op (rhymes with hisssss-op). Be kind to the sysop, as he can raise or lower your access level with the pressing of only a few keys.

SOMEONE TOLD ME TO SEND A LETTER VIA FIDONET. WHAT'S A FIDONET?

FidoNet is a network of bulletin boards all across the nation that exchange mail. They allow you to send mail to users on other BBSs! For example, from my home in San Diego I can send a letter to a friend who frequents a BBS in New York City. The BBS will automatically find the electronic path to the bulletin board in New York; the letter will bounce from system to system along the path until it finally arrives, usually about a week later. FidoNet is a fun toy, but its popularity is quickly dying out.

THE SYSOP OF A BBS I CALLED REQUESTED DONATIONS. WHY SHOULD I SEND ONE?

It's expensive to run a bulletin board—expenses such as electricity, phone costs, etc. have to be paid every month. Some system operators ask *regular* callers to send in contributions and help cover some of the expenses. To prove to users that any incoming money is used strictly on BBS expenses, some sysops ask you to make checks payable to the phone or electric company.

WHAT'S *YOUR* FAVORITE COMMUNICATIONS PROGRAM?

ProComm is my favorite for the IBM class of machines; *Red Ryder* is reputedly the best for the Macintosh family.

WHAT'S *YOUR* FAVORITE BBS SOFTWARE PACKAGE?

I'm partial to the ProLine and People-Net BBS programs. They're not as popular as the other programs, but they're at least ten times more powerful. The TBBS software is decent. I loath RBBS and WWIV.

APPENDICES

APPENDIX A

Telecommunications

Below are listed the most popular telecommunications software for the various operating systems. Remember, every communications software program is different, and each one has its own loyal followers insisting it alone is the greatest, most feature-laden software. Commercial software is often no better than the shareware communications software, so it pays to check out the shareware offerings first. The following programs are just a small sample of the software that is available for your computer. Good luck finding the perfect program for you, and happy modeming!

IBM/COMPATIBLE TELECOMMUNICATIONS SOFTWARE

- *Boyan* 4.0 ($55)
  Boyan Communications, 9458 Two Hills Court
  Columbia, MD 21045-3228 (919) 682-4225 (BBS)
- *Crosstalk XVI* 3.7 ($195)
  DCA/Crosstalk Communications, 1000 Holcomb Woods Pkwy.
  Roswell, CA 30076 (404) 998-3998, (404) 998-8048 (BBS)
- *Crosstalk Mk. 4* 1.1 ($245)
  DCA/Crosstalk Communications, 1000 Holcomb Woods Pkwy.
  Roswell, CA 30076 (404) 998-3998, (404) 641-1083 (BBS)
- *Freeway Advanced* 3.0 ($139)
  Kortek, Inc., 460 California Ave.
  Palo Alto, CA 94306 (800) 327-0310, (415) 327-4555

- *Mirror III* 1.0 ($99.95)
  SoftKlone Distrubuting Corp., 327 Office Plaza Dr., 100
  Tallahassee, FL 32301 (904) 878-8564, (904) 878-9884 (BBS)
- *ProComm Plus* 1.1b ($89)
  Datastorm Technologies, Inc., 3212 Lemonne Blvd., P.O. Box 1471
  Columbia, MO 65201 (314) 474-8461, (314) 474-8477 (BBS)
- *Qmodem SST* 4.0 ($45, manual $15)
  The Forbin Project, Inc., P.O. Box 702
  Cedar Falls, IA 50613 (319) 232-4516, (319) 233-6157 (BBS)
- *Relay Gold* 3.0 ($295)
  Microcom Software Division, 41 Kenosia Ave.
  Danbury, CT 06810-9990 (800) 847-3529, (203) 798-3800
- *Smartcomm III* 1.1a ($249)
  Hayes Microcomputer Products, Inc., P.O. Box 105203
  Atlanta, GA 30348 (404)-441-1617, (800) 847-2937 (BBS)

IBM/COMPATIBLE BULLETIN BOARD SOFTWARE

- *TBBS* ($299.95)
  eSoft, Inc.
  4100 S. Parker Rd. #305
  Aurora, CO 80014 (303) 699-6565
- *PC Board* ($120)
  Clark Development Co., Inc.
  P.O. Box 571365
  Salt Lake City, UT 84157-1365, (801) 261-1686
- *American EMail* ($149)
  Smith Micro Software
  P.O. Box 7137
  Huntington Beach, CA 92615 (714) 964-0412

CONNECTIVITY SOFTWARE

- *MacBlast* ($195)
  asynchronous file transfer software
  Mac Plus or larger
  Communications Research Group, 5615 Corporate Blvd.
  Baton Rouge, LA 70808 (504) 923-0888 or (800) 242-5278

- *MacLinkPlus* ($200)
  file tranfer and translation software
  Macintosh Plus or larger
  Data Viz, Inc., 35 Corporate Dr.
  Trumbull, CT 06611 203-268-0030
- *Laplink Plus* ($139.95)
  file transfer software
  IBM compatible desktops and laptops
  Traveling Software, Inc., 18702 North Creek Parkway
  Bothell, WA 98011 (800) 343-8080
- *ProLink* ($59.95)
  file transfer software
  Apple II and Macintosh
  Alsoft, Inc., PO Box 927
  Spring, TX 77383 (713) 353-4090

MACINTOSH COMMUNICATIONS SOFTWARE

- *MicroPhone II* ($295)
  telecommunications
  Macintosh 512KE or larger
  Software Ventures Corp., 2907 Claremont Ave., Ste. 220
  Berkeley, CA 94705 (800) 336-6477
- *Red Ryder* ($80)
  telecommunications
  Macintosh 512K and larger
  FreeSoft, 150 Hickory Dr.
  Beaver Falls, PA 15010 (412) 846-2700
- *Smartcom II* for the Mac ($149)
  terminal emulation
  Any Macintosh
  Hayes Microcomputer Products, Inc.
  PO Box 105203
  Atlanta, GA 30348 (404) 449-8791
- *Quick Link II* ($99)
  telecommunications
  Mac 512K and larger
  Smith Micro Software, Inc., PO Box 7137
  Huntington Beach, CA 92615 (714) 964-0412

MACINTOSH BULLETIN BOARD SOFTWARE

- Red Ryder Host ($135)
  Macintosh 512K and larger
  FreeSoft, 150 Hickory Dr.
  Beaver Falls, PA 15010 (412) 846-2700

ATARI COMMUNICATIONS SOFTWARE

Commercial

- *Flash* ($24)
  Antic, 544 Second Street
  San Francisco, CA 94107 (415) 957-0886
- *Interlink* ($32)
  Intersect Software, 2828 Clark Road
  Sarasota, FL 34231 (800) 826-0130

Public Domain

- *Uniterm*
  c/o Simon Poole
  Bahnhofstr.51B CH-5430
  Wettingen, Switzerland

AMIGA COMMUNICATIONS SOFTWARE

- *A-Talk III* ($99)
  telecommunications
  500, 1000, 2000
  1024k
  Oxxi, Inc.
- *Amiga Term* ($49.95)
  telecommunications
  500, 1000, 2000
  256k
  Commodore Business Machines
- *Diga!* ($79.95)
  telecommunications
  500, 1000, 2000
  512k
  Aegis Development, Inc.

AMIGA BULLETIN BOARD SOFTWARE

- *BBS-PC* ($149.95)
  500, 1000, 2000
  512k
  Micro Systems Software

APPLE COMMUNICATIONS SOFTWARE

- *Commworks* ($95)
  telecommunications
  All Apple IIs
  128k
  PBI Software
- *Point-to-Point* ($129)
  telecommunications
  All Apple IIs
  128k
  Pinpoint Publishing
- *Procomm-A* ($49)
  telecommunications
  All Apple IIs
  48k
  Prometheus Products Inc.
- *Talk is Cheap* ($40)
  telecommunications
  All Apple IIs
  128k
  Carolina System Software
- *Smartcomm I* ($119)
  telecommunications
  All Apple IIs
  128k
  Hayes Microcomputer Products

APPLE BULLETIN BOARD SOFTWARE

- *Bulletin Board Construction Set* ($49.95)
  All Apple IIs
  64k
  Sourceview Software Int'l.

- *Prime Bulletin Board System* ($99.95)
  All Apple II's
  64k
  Computorworks

COMMODORE COMMUNICATIONS SOFTWARE

- *Bob's Term Pro ($59.95)*
  telecommunications
  128
  64
  $49.95
  Progressive Peripherals and Software
- *Communicato*r ($200)
  telecommunications
  128 and 64
  Free Spirit Software, Inc.
  Ultra Terminal-64 (smart terminal package)
  128 and 64
  $49.95

COMMODORE BULLETIN BOARD SOFTWARE

- *C-Net* ($89.95)
  128 with RGB monitor
  64
  $64.95
  Softech

APPENDIX B

A Beginner's Guide to Installing a Modem by Jack Dunning

I'm not an expert on modems or telecommunications, though I have a pretty good understanding of what a computer can do. I have installed numerous modems on different types of computer systems and I can say confidently that it has never gone perfectly for me. I've made almost every mistake possible, which makes me imminently qualified to guide you through your first modem installation.

If you haven't purchased your modem yet, you may want to read this section first. It may save you time, money and trips to a computer store. Most of the computers I've worked with were IBM compatible, though I have worked with the Macintosh enough to be familiar with it. However, most of the approaches and ideas I offer will apply to almost any computer.

HAYES COMPATIBILITY

Most of the modems you will consider buying will be Hayes compatible. Hayes developed a language for computers to talk with modems. Modem manufacturers and software developers liked the language so much that it became a standard. This section assumes you either already own a Hayes-compatible modem or you are planning to buy one. If you own any other type of modem, I don't have a clue other than to tell you to read the manual.

EXTERNAL VS. INTERNAL

Your first decision is whether to get an external or internal modem. An external modem is in a separate box, comes with a power supply and generally has a bunch of lights on the front of the box. These lights help to solve problems of installation and communication, besides giving you the feeling of power while sending your data bits around the world. Be warned: The more lights and the better the box, the more expensive the modem. The box and lights make up about half the cost of the external modem.

There are more advantages to external modems than just lights. External modems are portable, which means they can be attached to almost any type of computer with few compatibility problems. Changing a hardware switch setting on the external modem is generally easier. Finally, it comes with a separate power supply—making it necessary to buy that power block with surge protection you've always wanted.

The internal modem is a Printed Circuit Board (PCB) that plugs into one of the expansion slots inside your computer. (Your computer must be expandable to host an internal modem.) Each internal modem is designed for a specific type of computer (i.e. IBM or compatible) and will only work in that type. Since it is hidden inside, it will take no additional space on your computer desk. There are no lights, but like the external modem, there is generally a speaker mounted on the board. The speaker is a tremendous aid in getting your modem up and running. To install an internal modem, you must open your computer and venture inside. Being dedicated and buried in the host machine, an internal modem is not very portable.

Advantages of the internal modem include no requirement for an outside power supply. It gets its power from the computer. There is no special cable required to hook it up to the computer.

(As you will see, this can save many headaches.) The internal modem does not require a serial port in the computer because it acts as its own serial port. Speaking of serial ports, it's time to clear this subject up.

SERIAL PORTS

Also know as asynchronous ports and RS-232 ports, serial ports serve as the first link between a computer and the telephone line. A serial port sends the data bits down one wire, in a row, making it possible for the data to go over a telephone line, which only has two wires; one to receive and one to send. For telephone communications with a computer, something has to act as a serial port. The modem communicates with the computer via a serial port. When two computers are communicating, the first computer sends data via its serial port to the modem. The modem in turn sends the data over the telephone wire to another modem on the other end. The other modem then sends the data to the receiving computer via its serial port. This is an important concept in installing and problem solving with your new modem.

Most IBM-compatible computers come with at least one serial port installed; some have two. Check your documentation to determine what you have in your computer. Some PCs will show you what is installed at startup. (Remember, it may be called an asynchronous or RS-232 port by your computer.) On ATs the setup program should tell you what ports you have.

If you don't have a serial port or if you are using your serial port for some other function, such as a graphics tablet or a mouse, you'll have to buy one, unless there is an internal modem available for your brand of computer. Remember, an internal modem has its own serial port, eliminating the need to use one that came with the computer. Tip: Before you buy an internal modem, make sure you have an empty expansion slot in your computer.

On the Apple Macintosh Plus and higher models there are at least two serial ports. One is marked with the phone icon and the other with the printer icon. Normally, the phone icon would be used for your modem. The communications software I used allowed me to select the desired serial port when I loaded it.

For most computers, your dealer should be able to tell you if you have a serial port, and if not, how to get one. Dealers can let you know if there are internal modems available for your computer. Assuming that you have solved the serial port problem, we will now solve the cable problem.

MODEM CABLES

If you intend to buy an internal modem, you'll have no cable problem because the modem is inside your machine. For all other situations you will have to buy a cable to run from your serial port to the modem. Modems don't come with cables because the modem manufacturer doesn't know what kind of computer you own. And it does make a difference!

In the case of Macintosh and most non-IBM compatible computers, your dealer should be able to supply the appropriate cable without a second thought. Each Mac uses the same connector on its back panel and each Hayes-compatible modem uses the same connector on its back panels. Any Apple dealer should carry modem cables for the Mac.

IBM and compatible computers are not so simple. You will need to know the exact specifications of your computer's and modem's connector to determine the cable you need. The modem end of the cable is easy since most modems are fairly standard. On that end the cable will have a 25-pin male connector that fits into the 25-pin female connector on the modem.

The concept of male and female connectors was invented by engineers who only thought about things two different ways. Something is either an engineering problem or a social problem. Getting the two problems confused, they determined that male connectors were the ones with prongs which were, well, uh ... you know. The plug on your vacuum cleaner is a male connector whereas the wall socket is a female connector. (These are the same engineers who wanted to name their kids after famous numbers.) Now it is time to identify the gender and type of the connector on the serial port on the back of your computer.

Once you're over your embarassment, look behind your computer at the back panel. Now try to identify your serial port. On name-brand clones such as Epson, the ports may actually be labeled. One will be parallel or printer, and the other should be serial, asynchronous or RS-232. Look at the serial port connector and determine gender and the number of pins or pinholes. Count the holes if necessary. Write it down! If you don't write it down, I guarantee that you will become confused by the time you're ready to buy the cable. The 25-pin connectors may be either male or female, depending on your computer. The nine-pin serial port connector found on most new PC compatibles and ATs is generally a female connector. (If you see a nine-pin male connector, it is probably your game port for a joystick.)

The parallel port may look exactly like the serial port, if both are 25-pin connectors and female. In this case the serial port is the one not plugged into your printer, if your printer works. On many of the clones the parallel port has a 25-pin female connector and the serial port has a 25-pin male connector.

Once you have identified your serial port, label it. Most IBM clones do not have labels on their ports. If you are unable to identify your serial port, call your dealer or your buddy who knows everything about computers. You must know the size and gender of your serial port connector to buy the right cable for your modem. It is better to make this determination now, than to later explain to your spouse why you have a dumb look on your face when you've discovered that the cable doesn't fit your computer.

Null Modem Cables

You probably do not want a null modem cable. A null modem cable switches the send and receive pins on one connector so tw computers can talk to each other without modems or phone lines. A modem is already set up to talk to most computers with a regular cable.

Are we lost yet? If you have a serial port and know which cable you need, we are ready to proceed.

THE PHONE LINE

To achieve your goal of on-line communications you are going to need a telephone line. All you need is a standard telephone line with the standard wall jack. (You must have a modular jack for your phone line.)

You can use your home phone line, but remember that during the long hours you're communicating with other computers you won't be able to receive other phone calls. "But," you say, "I have call waiting!" If you have call waiting in force, the obnoxious clicking sound that it makes will be enough to knock you off-line while you're in the middle of a computer conversation. (You'll want to cancel Call Waiting for the duration of the call by prefacing your dialing with *70 on touchtone service and 1170 on pulse service. You may have to add commas to your dial sequence to allow time for the dial tone to return.)

Some people have the phone company install a separate line to keep the regular home phone open for calls. You may want to wait to install another line until you decide how much you're going to use your modem. In any case, how far is your computer from the phone jack? The phone cord that comes with the modem is only six feet long. If your computer is more than six feet from the phone jack you have three choices: Move the computer next to the phone, install the phone closer to the computer, or get a longer telephone cord. Any of these options will work, but which one will your spouse tolerate?

Some people, especially phone company people, will tell you that you need a special, more expensive data line to use a modem. Don't you believe it! Unless you are doing special, super-high-speed transmission over a long distance, you don't need it. Besides, they don't put in a special line for a data line, they just test it to a higher quality level. If, after you get your modem connected, you find that you're having problems on your phone line, *then* you may want to consider a special data line.

COMMUNICATIONS SOFTWARE

Communication software is the piece of the puzzle required for completing the modem's installation and making sure it is working properly. This is the software that makes it possible for your computer to communicate via the serial port with your modem, ordering the modem to dial the phone and thereby calling another modem, which will awaken its host computer via its serial port, which in turn responds via its communications software.

Luckly, tons of communications software is available. Just look at the listings in the back of this book. And a lot of it is free. Many of the internal modems come with software (I think this is to make up for the fact that they don't have lights and a box). Some of the software is in the public domain, which means it's free if you can find it. Check with your buddy who knows everything about computers. Read reviews. Talk to your dealer. But get some communication software; you'll need it if we're ever going to complete this installation.

Hint: It would be worthwhile to get a software package with a real manual. About half of the problems with installing and running a modem for the first time are in the software. I'll be able to point you in the right direction, but some of those directions may require you to figure out how to do it with your particular software. Documentation helps, especially if there is an index.

CHECK LIST FOR MODEM INSTALLATION

Internal Modem

- Empty Expansion Slot
- Compatible Internal Modem
- Phone Line With Proximate Modular Jack
- Communications Software

External Modem

- Serial Port In Computer
- Cable With Proper Pin-Number and Gender at Both Ends
- Phone Line With Proximate Modulor Jack
- Communications Software

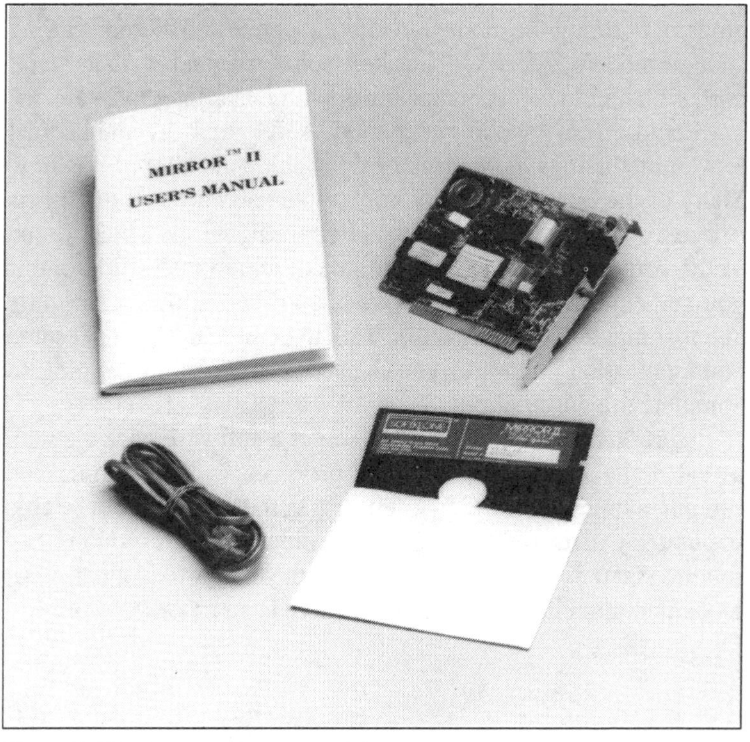

Figure 11: The above is a good example of what you receive when you buy an internal modem. Notice that there is no costly serial cable to deal with.

Figure 12: *When you buy an external modem it will not come with a serial cable. This photo illustrates a regular 6-foot serial cable with 25-pin male connectors on both ends. With the exception of the cable, this is what you receive when you buy an external modem.*

READY TO BEGIN INSTALLATION

Now that we've returned from our local computer dealer with all of the components discussed, we're ready to start. In preparation for this set of instructions, I installed both an internal modem and an external modem on an IBM-compatible computer, as well as an external modem on a Macintosh. Even if you don't have one of these computers, the procedures should be basically the same. And, who knows, maybe you'll make the same mistake I did.

The Modem Package

Open the box with the modem in it and look inside. For the external modem there should be a modem, a power supply, a six-foot phone code, a manual and a warranty registration card. The internal modem comes with the same, but replacing the modem box with a modem printed circuit board, and no power supply. Your modem may or may not come with a diskette that holds communications software, depending on the brand you buy. Don't fill out the registration card until you are positive the modem is working properly. Some dealers will refuse a return if the registration card is missing.

Carefully remove the manual. In spite of your tremendous desire to plug your modem in immediately and start playing (that's what I usually do), read, or at least peruse, the manual. This will help you to get familiar with your particular box—making it easier to solve problems later.

First you will find some words about one of our fine government agencies, the FCC. One manual told me that I must call the telephone company and tell them I'm putting a modem on my phone line giving the FCC registration number at the same time. Another one stated that it was not necessary to notify the telephone company, but that I better be ready to give information if the phone company asks for it. Being the daring type who likes to live on the edge, I didn't call the phone company. I know the only reason they want to know that I have a modem is so they will be able to call my computer in the middle of the night and suck vital data out of it. It must be a Communist conspiracy. Besides, they were on strike at the time anyway. So far, so good.

There are also instructions on what to do if your modem, heaven forbid, interferes with your television or radio. I've never seen this problem, but if you get it, the basic answer is move. Move the antenna, the computer, the television or to another state where they don't have these problems. What most modem users do is give up watching television. Actually, this is rarely a problem, but the government people need something to save us from.

Further in the modem's manual you will find the capabilities of the modem described, and a listing of commands from the Hayes modem language. Appendix C of this book contains the Hayes command set, for your convenience. You will also find switch settings that may vary for different software packages. These parts of the manual you will use for reference during the installation of your modem.

First I will describe the plugging in of an external modem, then the installation of an internal modem. Go directly to the section which is appropriate for you. Note: All installations of boards were accomplished on an IBM compatible computer, therefore the terminology and commands will refer to IBM PC and compatible systems. If you have a different type of computer, check your user's manual for appropriate commands and settings that correlate with the explanations given in this section.

PLUGGING IN AN EXTERNAL MODEM

If you already have a serial port which you intend to dedicate to your modem, the process will consist of plugging in the proper male connectors to the appropriate female connectors. If you need to add a serial port to your computer, the first thing to do is install that board.

Adding a PC Serial Port

If you're squeamish about opening your computer and looking inside, take it to your dealer and have the serial port installed. It probably isn't worth the hassle to do it yourself and besides, they have screw drivers and things like that. They will even make sure it works. But if you're the adventurous type, read on.

Turn off the computer and, to be safe, unplug it. Remove the case. (Check your computer manual for screw locations and how the case opens.) According to the instructions that come with the serial board, set the port to COM1, COM2, COM3 or COM4. COMn is a name that the computer uses to tell the different serial ports apart. It's important that each serial port have a different name. Otherwise the computer will get confused and no port will work. It's like two houses having the same address; the postal service wouldn't know where to deliver the mail, therefore they would probably give it to the neighbor down the street who in turn would throw it away.

If you already have a serial port being used for a mouse or something, it's probably COM1, so you would want to set the second port to COM2. If you're adding a first serial port, you should call it COM1. If you are changing the COMn setting on the board, you will probably either flip a tiny switch or move a jumper. (The drawings under "switch figure" will explain the difference in the switches and how to set them.) Set switch configurations according to settings listed in the serial board documentation. (If no documentation came with your serial port, call the dealer for this information.)

Putting it In

You should find some empty slots in the back of your computer. Remove the back plate next to the empty slot, insert the board into the connector and screw down the board with the same screw that held the back plate in place. You can now throw away the small back plate or use it as a tongue depressor. I now have a collection of about 50 back plates from various computers in which I installed miscellaneous expansion boards. Someday it will be almost as good as collecting stamps.

If, to your horror, you find that there are no empty expansion slots—due to either limited expansion space or your computer suffering from expansionitis (too many neat things inside your computer), you may still be in luck. Go back to your dealer and get a multi-function board to replace some of the functions now hogging a slot in your computer. I'm sure there is one that will fit your need. Put the cover back on the computer and we're ready to move on.

The Serial Cable

Now we take our carefully selected cable and plug one end into the serial port and the other end into the modem. If it doesn't fit, go back to the section on cables, then back to your dealer. Also, explain to your spouse why you're stomping around the house and swearing.

Plug in the Power Supply.

For most this would be simple, but as you look at your wall sockets you realize that one socket is being used by your computer and the other is being used by your printer. (Your monitor is plugged into the back of your computer, not requiring a wall socket.) There's no place to plug in your modem! It's time to go down to the local electrical supply store and buy a power strip so you can plug in more than two pieces of equipment. Make sure you get the one with the three-prong plug for grounding. Also, explain to your spouse the reason for your frustration.

Now that you're plugged into the computer and the power supply, skip past the section, "Installing an Internal Modem" to "Plugging in the Phone Line." From there on, the procedures for getting modems running are very similar.

INSTALLING AN INTERNAL MODEM

If you're squeamish about opening your computer and looking inside, take it to your dealer and have the modem installed. It probably isn't worth the hassle to do it yourself and besides, they have screw drivers and things like that. They will even make sure it works. But if you're the adventurous type, read on.

Turn off the computer and, to be safe, unplug it. Remove the case. (Check your computer manual for screw locations and how the case opens.) According to the instructions which came with the modem board, set the port to COM1, COM2, COM3 or COM4. "COMn" is a name the computer uses to tell the different serial ports apart. It's important that each serial port have a different name. Otherwise the computer will get confused and no port will work. It's like twins with the same name; if they get a call and you answer the phone, you wouldn't know with which twin the caller wanted to speak. Therefore, you would probably hang up the phone rather than go through this ordeal again.

When I installed the internal modem, I already had one serial port in the computer, COM1. Therefore, I wanted the modem to be COM2. I checked the manual, looked at the jumpers on the board (See Switch Types), and found that it was already configured for COM2. I figured the modem's manufacturers realized most PC compatibles were configured with a serial port called COM1, so they configured the modem for COM2. Just for the heck of it, I reconfigured it for COM3. I wanted to see if it would work. (And it did.)

Check your manual and set the jumpers or switch to the appropriate selection for your situation. If you are using slot eight on the IBM PC-XT, there may be a special setting on the board. Once you have the modem board set up for your situation, follow the directions explained in "Pluggin It In," under "Adding a PC Serial Port," in the "Plugging In an External Modem" section. The procedures are the same.

PLUGGING IN THE PHONE LINE

Next, plug the phone cord into the modular telephone wall socket and the other into the modem. You may notice that there are two phone sockets on the modem. (If there's only one, your choices are narrowed.) And, yes, it does matter which one you plug the phone line into. One socket on the modem is for the phone line and the other is for a telephone. Having a telephone attached to your modem allows you to quickly pick up the receiver and appologize when you've dialed what you thought was a BBS number at three o'clock in the morning and actually get a little old lady (and she had just finally fallen asleep at two o'clock in the morning).

On some modems, there will be a label or icon to tell you which socket to use. For other modems, only reading the manual will give you a clue. Locate the appropriate socket and plug in the phone line.

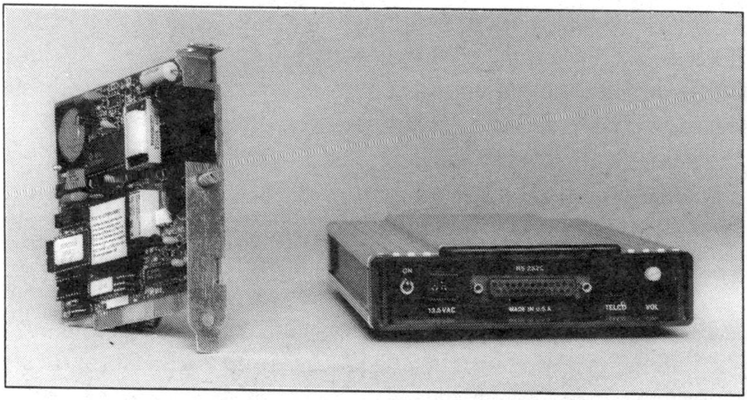

Figure 13: *The internal modem above (the one with the rather blank-looking back) has two modulor phone jacks and a volume control. The external modem is a bit more detailed with a serial connector, A/C jack, volume control and two phone jacks.*

Plug in your computer and turn it on. External modem users, turn on your modem. You will notice at least a power light, maybe more, light up depending on how many lights your modem has.

As we go through the problems of getting the modem running, I will describe the difference between what the internal modem user and the external modem user will probably see and hear. As a rule, both will see the same things on the computer screen and hear the same sounds from the modem. The sounds from the internal modem will be just a little more muffled since the noise must pass through the computer case. Another difference is the internal modem does not have any of the indicator lights that are found on the external modem, which at times may make it more difficult to determine exactly what's happening.

LOADING THE COMMUNICATIONS SOFTWARE

If you have previously read or at least scanned your communications software manual, this entire procedure will be much easier. If you haven't done that yet, take the time now to do so.

The external modem I installed actually came with a list of various popular communications software and the corresponding DIP switch on the modem for optimum operation of each (these switches control features as auto-answer, enable/disable, etc.). If yours doesn't have such a listing, don't worry; the factory setting generally will work just fine with most software. Modems are more forgiving than most computer equipment. Almost any of the settings can be overridden by the communications software. If your modem doesn't appear to have any switches other than the power switch, it would be rare that you would have to change them. The internal modem I installed had no hardware switches for these features.

Load Your Communications Software.

If this is the first time you are loading the package, many programs will run an installation routine. If no installation routine is included, there are a number of parameters you will want to check or set before you actually try to dial a number with your modem.

First, check that you are set to use the proper serial port (i.e. COM1, COM2, COM3, etc.). On the Macintosh, the software prompted me to select the port with phone icon or the print icon on start-up. Then, check the baud rate to match that of your modem. Set it to the highest speed available on your modem, usually 1200 or 2400 bps. If you hook up to a slower modem, your modem will automatically adjust down.

There are other parameters such as bits, stop bits, parity and duplex. Set them to eight bits, one stop bit, no parity (none) and full duplex. There are times when you may want to change these for communication with special systems, but most of the time these settings will work.

Some programs, upon loading, will go through a procedure called "initializing the modem," where checks will be done with the modem to ensure communication. If the modem doesn't respond, the software will prompt you with "modem not responding." This could be something as simple as the modem being turned off or the wrong serial port being selected. If you get this message, recheck your setup and reinitailize (there is usually a separate command for reinitialization). When doing an intialization routine on an external modem modem you will be able to see the RD (Receive Data) and SD (Send Data) lights lighting up. Communications software doesn't necessarily have an initialization routine, but if yours does you'll find it helpful when your ready to go on-line.

If you have an external modem, the TR should come on when you have loaded your software and the computer has found the modem. TR stands for Terminal Ready, indicating that both the computer and modem are communicating and ready to go.

THE FIRST TEST

Get a phone number to call. In many geographic areas there are listings of on-line computer systems you can call for free. If you have a local computer magazine, more than likely they will list some local numbers. If not, talk to user groups, computer dealers or your buddy who knows everything about computers. One of them will get you a number. (You can do some of your testing without an actual on-line host system, but you won't be able to get communications.)

Now dial! With some software there is a single keystroke to initiate, then you are prompted for the number you would like to dial. Others use the Hayes command set directly, which would be "ATDT ###-####." (###-#### represents the phone number.) AT stands for Attention to the modem. Almost all Hayes commands must be preceded AT. DT stands for Dial Tone, which will instruct the modem to dial using tones of the number that follows. If you want pause in the number you should insert commas.

Once you have executed the dial command the following should happen:

- You will hear a click like the phone receiver is being lifted over the speaker. On the external modem the OH (Off Hook) light will come on.
- You will hear a dial tone.
- You will hear the tone dialing the number you entered. If the phone line doesn't respond to the dialing, it may be because you have the older pulse type dialing system. In that case you will want to change your dialing parameter to pulse. Your software may have a separate installation utility to change. The Hayes command would be "ATDP ###-####." When you are using pulse dialing you will hear the clicking sounds, just like that of the rotary phones.
- You will hear the ringing of the phone. Your heart starts beating faster in anticipation of your first on-line experience. Or, you hear a busy signal and realize that you will either have to call back or get another number. In either case, at this moment you know that both your software and modem are working, at least up to this point. Normally, if you hit the return key at this point, the modem will hang up the phone. At worst, the command ATH will hang up the line.
- The modem on the other end of the line answers the phone and you hear a high-pitched, continuous tone. This tone is

called "carrier" and is used by the modem to establish the communication link.

- When your modem detects the carrier, it will respond with another high-pitched, continuous tone. On an external modem, the CD light will be turned on indicating Carrier Detected.
- About a second or less later, you should hear the "kshchshchshch" sound. (This is a noise that can be made by clenching your teeth together, pushing the middle of your tongue against the roof of your mouth while pressing the tip of your tongue against your inside gums just below your lower teeth, then forcing air through your cheeks and out your open mouth. It sounds like a water hose on high speed.) Connection has been made. The two modems are communicating. All you have to do now is respond to the host computer.
- After another second the speaker will turn off, because you no longer need it. You should see a sign-on message from the host system. If not, try hitting the return key. Sometimes that is all it takes to get the host system going. On an external modem, when you strike a key you should see the SD light flash to indicate that you are sending data. Conversely, you should see the RD light flash when you are receiving data.

If all these things happened as I described, you're ready to go and you can put this book down. If everything happened but the host system doesn't seem to be responding, don't assume it is your fault. Try another number for another system. There are numerous reasons why a host system may hook up with you, yet show no further response and most of them have to do with the host.

TROUBLE SHOOTING

I Entered the Dialing Command With a Number, but I Never Heard a Thing

If your modem has a speaker, and not all modems do, it will also have a volume switch. Your volume may be turned down. If you have an external modem and the OH light comes on, this is almost certainly the problem. Find the volume switch and turn it all the way in one direction. Try dialing again. If you still hear nothing, turn the switch all the way in the opposite direction and try once more.

There is also a Hayes command that turns the speaker on and off. If you accidently sent the turn off command, ATM0, then you will want to send the original defaults ATM1 for normal operations, or ATM2 to leave the speaker on continuously. If there is still nothing, you have a real problem and must resort to other tests.

We don't know whether you have a software problem, a connection problem or a modem problem. To isolate the problem, we will first remove the software as a variable. (This procedure is designed for the IBM PC and compatibles. I have not done a similar procedure with other types of computers, but I'm sure it is possible. The principles are the same.)

Leave the communications software and return to the operating system level, DOS. Use the mode command to configure your serial port by entering: MODE COM1:2400. The MODE command is a DOS program that will configure your serial port to match the baud rate of the modem. You should choose the port (COM1, COM2, etc.) according to where you think you have your modem installed.

Then, with your word processor or a text editor such as EDLIN, create a short non-document (ASCII) file with the command "ATDT ###-####." You can name this file TEST1. Now, send this file to the modem with the command COPY TEST1 COM1. The modem should get a dial tone and dial the number.

(You should be able to do this same test by redirecting your print with: MODE LPT1:=COM1, CONTROL-P, and sending the "ATDT..." command directly.) If it works, the problem is in your communications software and you will have to work with the manual. If it doesn't work there are a few more things you can try. (To stop the ringing or disconnect, you can either reset the modem by turning it off and on again, or send the command ATH to the modem via the same method.)

Your modem may be assigned to the wrong serial port, especial an internal modem. Send the same test file to COM2, COM3 and COM4 with the COPY TEST1 COMn command.

If none of these work, check your cables to make sure they are tight. Check your phone line to ensure it is in the right jack. Get a new phone cable and try it. If possible, you may also want to swap your serial cable. Cable problems are not unusual.

Some external modems have a switch that will reverse pins two and three in the serial interface. Some computers may require this. Pins two and three are the send and receive signals. Some computers, especially older types, have their serial ports set up with pins two and three reversed from what is expected by most modems. If this is the case and your modem doesn't have a hardware switch for reversing the pins, get a null modem cable. That is essentially what a null modem cable accomplishes.

Nothing has worked. At this point, there is a problem in either your serial port or modem. For an external modem check your switch setting against the manual. I would even systematically try different settings. It's not likely this will make much difference, but it's worth a shot and it won't hurt anything.

Now is the time to take your modem back to the dealer. If it turns out there is nothing wrong with the modem, the computer must go in. I once owned a computer in which a serial port just wouldn't operate properly. That's extremely rare, but it does happen. If your computer is still under warranty, your dealer should take care of you with no charge.

I Get a Dial Tone but That's All

If you don't enter a number after the ATDT, the modem will give you a dial tone and just sit there. If you entered the number and heard the tones dialing, but you still had a dial tone when it was finished, you are probably using tone dialing when you need pulse dialing, or vice versa. The pulse dialing command is ATDP.

If it still isn't working, try some of the tests in the previous question and take the modem back if you get negative results.

If you're accessing another line that gives you a new dial tone, your modem may be dialing before the new tone appears. To put in the appropriate pause, add commas to the dial string. Enough commas will cause the modem to pause long enough to achieve a new dial tone.

My Modem Indicates CD (Carrier Detect) Before I've Even Started Dialing

It is likely that your Carrier Detect switch is in the override position. Check your software manual and adjust accordingly.

I Don't Seem to Establish a Connection Even Though I Can Hear Both Carrier Signals

If you don't hear the "kshchshch" sound, your signals weren't able to synch. Often this is due to a difference in baud rates, but today most modems are so smart this is rarely the problem. Also, your parameters, as long as the word length is seven or eight bits, are rarely the problem. Sometimes two particular systems just have trouble hooking up due to slight differences in tolerances.

Try calling some other systems to see if you still have the problem. If you can't establish communications with anyone, the problem is definitely on your end. You might try a different modem. If you can get on with some other systems, but still not the first one, then try another speed, usually slower. Sometimes this can make the difference. If that fails, forget that system for now. It is very possible that the problem is at the other end.

I Connect, but All I Seem to Get Is Garbage

There are two types of garbage, random garbage and garbage with a purpose. Random garbage will give you odd characters ()]*&) with no apparent pattern. The characters will tend to keep coming, even when you're supposed to be answering questions. Also, there may be groups of words coming from the host system that you can actually read. This kind of garbage is most commonly caused by a bad telephone line.

I was lucky enough to have this happen to me during one of my installations. I disconnected the line from the modem and plugged in a regular telephone. (If you already have your phone connected, just pick up the receiver.) I could hear a snap, crackel, pop on the line. The static on the line was causing the noise.

I called the phone company, who was on strike at the time, and they told me that they would check it out. They have the capability to check out a line without even leaving the office. Later in the day, they called me back and confirmed that there was a problem on the line. They would send someone out to fix it tomorrow, but if they found that the problem was in the wiring in my house, I would be charged $35. And that wouldn't even fix my problem.

I asked them if I paid someone to check out my wiring and found that the problem was with the phone company, could I charge the phone company $35. Of course they said "no," and explained in detail that they didn't make the rules, they just worked there. Who makes these rules?

In any case, after a little more discussion, I went outside and disconnected the phone line at the terminal on the side of the house. (This is the same place where the mad killer always clips the telephone line in those horror movies.) When I called the technician back, in about 30 seconds he was able to determine that indeed it was a phone company problem and that they would send someone out the next day. I promptly apologized for being so irate earlier. After all, he was just a poor plebe with no control over his destiny, working in a job devoid of any future prospects and, who didn't deserve any abuse from me.

About an hour later, there was a knock at my door. It was a repair person from the phone company. I couldn't believe it, same-day service. He had to climb some poles, but was unable to immediately resolve the problem. He worked until six, but concluded that he would have to return the next day. Sometime the next day he solved the problem. I don't know when he actually finished because the next day was Monday and I had to go to work myself. I've never had such good service—and on Sunday, while the phone company was on strike.

So, if you're getting random garbage, there is a good chance that the problem is in your phone line. Sometimes going to a slower baud rate will solve your problem. I have experienced difficulties at 2400 bps on long distance calls and eliminated those by switching to 1200 bps. Naturally, slower speeds will make your phone bill higher.

Some people seem to live in areas where the quality of the phone lines just isn't as good as in other areas. If this seems to be the case for you, for a little more money you can get a special data line. The phone company will test the data line to a higher quality level, and if it is below specifications they will work to bring it up.

If the garbage you're getting starts and stops at regular intervals, you're getting garbage with a purpose. You won't receive even a little bit which is readable. This means that the host system is operating with a different set of parameters, causing your system to see things differently. You should be able to change your parameters while you're still connected via your communication software. If you're set to an eight-bit word length, one stop bit and no parity, try a seven-bit word length, one stop bit and odd parity. These are the two most common setups in my experience. One, or both, of these usually works.

I'll Be On-line and Working With Another System, When I'm Suddenly Off-Line. What's Happening?

There are a couple of possiblities here. The first is: Do you have call waiting? Call waiting service from the phone company causes an audible click on the line when there is an incoming call. This click is often enough to knock you off-line. If you think this may be the problem, you should disable call waiting before you make your call. To disable call waiting for the duration of the call, dial *70 on touchtone service and 1170 on pulse service prior to dialing your number. You may have to add commas to your dial sequence to allow time for the dial tone to return.

The other primary reason for getting knocked off-line is that the person on the other end doesn't want you there. There are two basic ways this may happen. One is that they log you off. If they are doing this, you will notice that the computer is magically making menu selections for you and those selections just happen to be logging you off. The sysop (System Operator) on the host system is making these selection for you for his or her own reasons. Most bulletin board systems have this capability.

The other ways the sysop can quickly dump you is to turn the host modem off and on again. Bingo, you're gone! If this is happening to you, there is nothing you can do except clean up your act, or change to a better deordorant.

I Keep Getting Double Characters. What Can I Do?

Most systems run in full duplex. That is, when you send a character to the host system it returns the same character to you, which is then displayed on your screen. The characters you see are actually originating at your keyboard, speeding out the serial port through your modem to the host system and back before they are displayed on your screen. In half-duplex mode your system is not expecting a return character, therefore it will be sent directly to the screen. If your system is set to half duplex and the host system is set to full, you will get double characters when you type. Just set your system to full duplex and your problem should be solved.

Is There Any Way That the Speaker Will Stay on After I Connect?

Yes. The Hayes command "Mn" will do that. Just send to the modem ATM2 and the speaker will stay on continuously. ATM1 will return it to the original default and ATM0 will turn off the speaker for good.

I Seem to Be Getting Outside Interference

It is possible for televisions, radios and other electrical devices to interfere with your communications. In general, you will want to move away from the offending device. If this doesn't work, put RF filters on all your power lines and make sure you use shielded cables. If you're still having problems, build a metal screen room to protect your entire system from RF radiation. If that still doesn't work, build a metal bomb shelter 50 feet beneath the ground with a dedicated power supply. Or, just take up some other hobby.

Are There Any Other Tests I Can Perform?

Yes, on some modems there is an Analog Loopback Self-test. This test checks the modem transmitter and receiver for proper operation. If your modem has this capability, the manual will so state and give you instructions on how to do the test.

Switch Types—The Differences and How to Set Them

The two types of switches you are likely to encounter are DIP switches and Jumpers. DIP switches are a row of tiny switches, each being about the size of the head of your pen. On some external modems they are on the bottom, fully exposed and labeled. This is the best situation because access is easy. To change a setting you merely take your pen, a small screwdriver, toothpick or some other small device and flip the switch to the desired position.

On some external modems you will find the DIP switches behind the face plate, requiring you to remove that plate before changing the settings. On others the switches are inside the box, indicating that the manufacturer doesn't want you flipping switches. Your modem manual is your best guide to switches.

The external modem uses shorting jumpers, located on the printed circuit board. The jumper posts look like a tiny row of two poles, with a occasional pair hooked together by a shorting jumper (see illustration). The selection of which pair of posts are hooked together determines the setting for the boards.

To change the setting use needlenose pliers or a pair of tweezers to carefully remove the jumper to be changed. Then slide the jumper onto the new posts at the desired position. Once you have done that you have become an expert on setting internal modems and can add this to your resume.

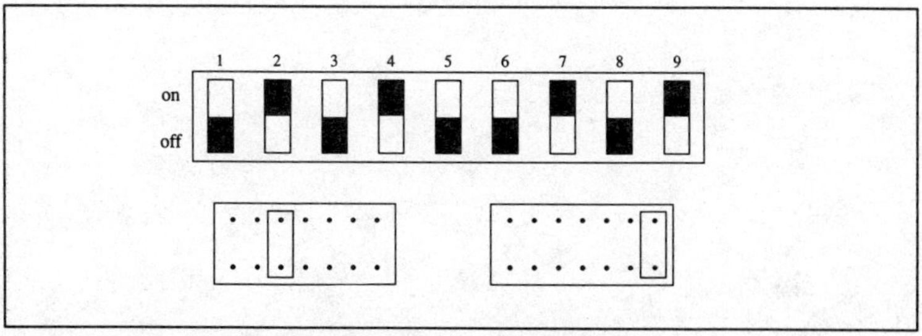

Figure 14: ***In the above illustration, there is an example of a row of DIP switches (on top) and a jumper in two different positions (on the bottom).***

APPENDIX C

Basic Command Set

Since Hayes is basically the standard when it comes to telecommunications interfacing, this appendix will give you a good idea of what the basic commands are. Any Hayes or Hayes compatible modem will be able to accept these commands. For further detail and a look at the extended set of commands, you can look in the user manual for any modem.

COMMAND	OPTIONS/FUNCTION
A	Force Answer Mode when modem hasn't received an incoming call.
A/	Re-execute the last command.
A>	Re-execute the last command (redial) continuously.
Any key	Terminate current dialing operation resulting from issued Dial command.
AT	Attention: must precede all other commands, except A/, A> and + + +.
B*n*	U.S./CCITT answer sequence. BO CCITT answer sequence B1 U.S., Canada answer tone—Default.

COMMAND	OPTIONS/FUNCTION
C*n*	Transmitter enabled/disabled. C0 Transmitter disabled. C1 Transmitter enabled—Default.
D*n*	Dial the number (n) that follows and go into Originate Mode. Use any of the following options: P Pulse dial—Default. T Touch-Tone dial. , (Comma) Pause for 2 seconds. ; Return to command state after dialing. " Dial the letters that follow. ! Flash switch-hook to transfer call. W Wait for second dial tone (if X3 or higher is set). @ Wait for an answer (if X3 or higher is set). R Reverse frequencies.
DS*n*	Dial number (n = 0—3) stored in NRAM.
E*n*	Command Mode local echo: display copy of modem commands entered at the keyboard. (Not applicable once a connection has been made.) E0 Echo OFF. E1 Echo ON.
F*n*	Local echo ON/OFF once a connection has been established. FO Echo ON. F1 Echo OFF—Default.

COMMAND	OPTIONS/FUNCTION
H*n*	On/off hook control. H0 Hang up (go on hook)—Default. H1 Go off hook.
I*n*	Inquirv. I0 Return product code. I1 Return memory (ROM) checksum. I2 Run memory (RAM) test. I3 Return call duration/real time. I4 Return current modem settings. I5 Return NRAM settings.
K*n*	Modem clock operation. K0 At AT13, display call duration—Default. K1 At AT13, display real time; set clock with AT13 = HH:MM:SSK1.
M*n*	Monitor (speaker) control. M0 Speaker always OFF. M1 Speaker ON until carrier is established—Default. M2 Speaker always ON. M3 Speaker ON after last digit dialed, OFF at carrier detect.
O	Return on-line after command execution.
P	Pulse dial.
Q*n*	Quiet Mode: result codes displayed/suppressed. Q0 Result codes displayed.

COMMAND	OPTIONS/FUNCTION
	Q1 Result codes suppressed.
Sr=*n*	Set Register commands: *r* is any. S-register, *n* must be a decimal number between 0 and 255.
Sr?	Query S-register *r*.
T	Tone dial.
V*n*	Verbal/Numeric result codes.
	V0 Numeric Mode.
	V1 Verbal Mode.
X*n*	Result Code options. Use the Options Table on this reference card. Default is X1, Extended set, codes 0-5, 10.
Z	Reset modem to defaults.
+++	Escape code sequence, preceded and followed by at least one second of no data transmission.
>	Repeat command continuously or up to 10 dial attempts. (Cancel by pressing any key.)
/	(Slash) Pause for 125 msec.
$	Help command summary.
&$	Help Extended command summary.
D$	Help Dial command summary.
S$	Help S-register summary.

APPENDIX D

Glossary

ANALOG LOOPBACK
This is a way of testing the functionality of your modem without the use of a phone line. By typing the command "AT&T1" (with Hayes compatible modems) all of the data sent to your modem is then "looped back" to the screen rather than going out through the phone line.

ANALOG SIGNAL
An electronic signal that is measured by way of some physical variable, such as pressure, voltage, or length. Analog signals are usually measured by their wavelength.

ASCII
American Standard Code for Information Interchange. ASCII data always consists of a standard seven-bit code with one parity bit. Otherwise known as "text," ASCII data can be transferred to and from most type of computer.

ASYNCHRONOUS
A form of data transmission otherwise known as serial. Asynchronous transmission allows data to be sent at irregular intervals. Each data unit is preceded with a start bit and followed by a stop bit. This is the common form of serial data transmission.

AUTO ANSWER
(AA) A mode that will allow a modem to automatically answer an incoming call without interaction by the user.

AUTO DIAL
A feature that some communications software programs provide for storing phone numbers in order to retrieve them later and dial them by executing one command.

BAUD
A unit of measurement that denotes the number of bits that can be transmitted per second. For example, if a modem is rated at 1200 baud it is capable of transmitting data at a rate of 1200 bits per second. The term was derived from the name of J.M.E. Baudot, a French pioneer in the field of printing telegraphy.

BBS
Bulletin Board System. A computer software application designed to answer calls and provide a message base for modem users. Usually used to store software and messages for access by anyone with a modem.

BIT
A unit of measurement that represents one figure or character of data. The word is derived from Binary digIT. A bit is the smallest unit of storage in a computer. Since computers actually read 0s and 1s each of these is measured as a bit. The letter "A" consists of 8 bits which amounts to one byte. See Byte.

BPS
Bits Per Second. Another way of measuring the speed of data transmission. See Baud.

BYTE
A group of consecutive bits of data. With most computers today a byte consists of 8 bits.

CPS
Characters Per Second. A form of measuring data output speed, much like baud or bits per second. This form of measurement is more commonly used to judge the speed of printers.

DIRECT CONNECTION
By using a null modem cable it is possible to directly connect one computer to another via their serial ports. By using communications software these two computers can communicate as if they were using modems. See Null Modem.

DUPLEX
A communication channel that allows data to be sent and received at the same time.

ECHO (Local, Remote)
Having a local echo ON will cause all transmitted data to be sent to the screen of the sending computer. A remote echo is such that everything that the remote computer (the one you are communicating with) transmits will be duplicated on your computer's screen.

FULL DUPLEX
Data is flowing in both directions at the same time. When Remote echo is ON communication is occurring in full duplex.

HALF DUPLEX
Data will flow in both directions, but only one way at a time. For example, when local echo is ON communication is occurring in half duplex.

HAYES COMPATIBLE
A modem is considered Hayes compatible if it supports the Hayes command set— a language used to communicate with and control a modem. See Appendix C.

HIGH SPEED (HS)
External modems with the capability of supporting more than one baud rate will illuminate the HS light when set on its fastest mode.

MODEM READY (MR)
External modems will have this light. The light will illuminate when the power is turned on. At this point the modem is "ready" for receiving commands, calls, or verbal abuse.

CARRIER
A term used to identify a signal of continuous frequency. Carrier is achieved when one modem recognizes the signal of another modem. If modem A is attempting to connect with modem B and there is no response, then modem A will find "no carrier."

CARRIER DETECT (CD)
This means that your modem has detected the carrier of another modem. Most external modems have a light for this on the front panel.

COM1,COM2, etc.
Most serial ports and internal modems can be configured to either COM1 or COM2 in order to accommodate the situation where both may exist. The DOS MODE command is used to change the output direction to such serial devices as modems.

DIP SWITCH
Dual Interface Poll switches allow for either an ON or OFF setting with any number of circuits. DIP switches commonly allow you to change the configuration of a circuit board to suit your particular computer.

EXTERNAL MODEM
A telecommunication device built into a printed circuit board and mounted in its own casing. This can be linked to a computer through the computer's serial port.

INTERNAL MODEM
A telecommunication device built into a printed circuit board to be used with a standard 8 bit expansion slot.

JUMPER
A jumper is a small block used to connect two terminal posts. Jumpers complete a circuit when attached to two poles and leave an open circuit when unattached.

NULL MODEM
A null modem is not a modem. It is actually a cable that is used to directly connect two computers via their serial ports. A null modem cable differs from a regular modem cable in that the transmitting and receiving pins are swapped.

ON/OFF HOOK (OH)
A modem is On Hook when it is sitting idle, neither sending nor receiving data. A modem is Off Hook when it is either sending or receiving. When you pick up a phone to dial a number the phone is Off Hook. All external modems will have a light for this.

ON-LINE
A term used for describing the status of a connection between two computers with modems. On-line means there is a connection. An on-line system is another name for a BBS. (see BBS)

PARITY BIT
A check bit used to make the sum of the bits in a unit of data either even or odd (including the parity bit). A unit of data that is 8 bits long would have no parity, and a unit of data 7 bits long would have an even parity bit to make an 8 bit word. Parity is used to check a unit of data for errors during transmission through phone lines or null modem cables.

PROTOCOL
When data is being transmitted between two or more devices

something needs to govern the controls that keep this data intact. Different protocols have been developed to try and come up with better and faster ways to transfer data between two or more computers.

RECEIVE DATA (RS)
External modems always have an RS light which will illuminate when data is being received from another computer.

SEND DATA (SD)
External modems always have an SD light which will illuminate when data is being sent through the modem over the phone line.

SERIAL CABLE
A cable used to connect peripheral devices through a computer's serial port. Normally a 25-pin connector on each end, yet can be a 9-pin on one.

SERIAL PORT
A serial port can either be plugged into an expansion slot on the motherboard of your computer or built into the motherboard itself. Serial ports are used for such devices as printers, mice, light pens, and, of course, modems.

START/STOP BITS
A start bit signals the start of a unit of data in asynchronous communications. A stop bit signals the stop of a unit of data. The unit can vary in length depending on the protocol.

TERMINAL MODE
Many communications programs allow you to mimic a computer terminal, which is basically a keyboard and CRT display and/or a printer. A common terminal mode emulator is VT-100.

TERMINAL READY (TR)
This light is illuminated when your computer has turned on the

RS-232 (serial) interface. Sometimes this light will not come on until you have loaded your communications software into memory.

INDEX

Other computer books from
Computer Publishing Enterprises:

How to Understand and Buy Computers
by Dan Gookin

How to Understand and Find Software
by Wally Wang

Parent's Guide to Educational Software and Computers
by Lynn Stewart and Toni Michael

Ten (& More) Interesting Uses for Your Home Computer
by Tina Berke

101 Computer Business Ideas
by Wally Wang

How to Start a Business With a Computer
by Jack Dunning

The Best FREE Time-Saving Utilities for the PC
by Wally Wang